D1319212

CHEMICAL PRINCIPLES IN THE LABORATORY

USING THE INTERNATIONAL SYSTEM OF UNITS

EMIL J. SLOWINSKI

Professor of Chemistry
Macalester College, St. Paul, Minnesota

WILLIAM L. MASTERTON

Professor of Chemistry
University of Connecticut, Storrs, Connecticut

WAYNE C. WOLSEY

Associate Professor of Chemistry
Macalester College, St. Paul, Minnesota

SAUNDERS GOLDEN SUNBURST SERIES

1979
W. B. SAUNDERS COMPANY • Philadelphia • London • Toronto

W. B. Saunders Company: West Washington Square
Philadelphia, PA   19105

1 St. Anne's Road
Eastbourne, East Sussex BN21 3UN, England

1 Goldthorne Avenue
Toronto, Ontario M8Z 5T9, Canada

Listed here is the latest translated edition of this book together with the language of the translation and the publisher.

German *(2nd Edition)*—Gustav Fischer Verlag, Stuttgart, Germany

CHEMICAL PRINCIPLES IN THE LABORATORY
Using the International System of Units   ISBN   0-7216-8383-5

Last digit is the print number: 9   8   7   6   5   4   3   2   1

PREFACE

In spite of its many successful theories, chemistry remains, and probably always will remain, an experimental science. Most of the research in chemistry, both in the universities and in industry, is done in the laboratory rather than in the office or computing room, and it behooves the young student of chemistry to devote a substantial portion of his time to the experimental aspects of the subject. It is not easy to become a good experimentalist, and it must be admitted that some famous chemists are not especially effective in the laboratory. Yet even those chemists who leave the actual work in the laboratory to a graduate student or an assistant must be completely familiar with the available experimental methods, the proper design of experiments, and the interpretation of experimental results. As beginning chemists, students will find that their efforts in the laboratory will be rewarded by a better understanding of the concepts of chemistry as well as an appreciation of what is required in the way of technique and interpretation if one is to be able to find or demonstrate any chemically significant relations.

In writing this manual the authors have attempted to illustrate many of the established principles of chemistry with experiments that are as interesting and challenging as possible. For the most part the experimental procedures and methods for calculation of data are described in great detail, so that students of widely varying backgrounds and abilities will be able to see how to perform the experiments properly and how to interpret them. Many of the experiments in this manual have not been published previously and were developed and tested in the general chemistry laboratories at the University of Connecticut and at Macalester College. We have included more experiments than can be conveniently done in the usual laboratory program, so that instructors may select experiments in a flexible way to meet the needs of their particular courses. In many of the experiments unknowns are to be assigned to students to ensure their working independently and to introduce a measure of realism to the application of the chemical principle being investigated.

It is always difficult to get students in chemistry to prepare in advance for the laboratory sessions. If this is not done, a substantial amount of the laboratory period will be spent in trying to find out what the experiment is about and how to do it. It may result in the student not finishing the problem or having to take the report with him to complete it. In order to encourage preparation for the laboratory, we have included with each experiment an advance study assignment, which includes a few questions that require the student to read the experiment and understand it to the point where he can make calculations if given appropriate data. Students who complete the advance study assignments will be better able both to do the experiment properly and to complete it in the allotted time.

In preparing this SI version of the lab manual we have used the same experiments as appear in the revised version of the second edition. All data in the experiments and Advance Study Assignments have simply been converted to SI units. The choice and order of the experiments are compatible with the SI version of our text, "Chemical Principles," but might be appropriate for use with any general chemistry text employing SI units.

iii

The authors would like to acknowledge the assistance of Miss Patricia Thiel of Macalester College, who tested the experimental procedures in the new experiments for clarity and feasibility and was very helpful in many ways in preparing the manuscript, and we appreciate the very willing assistance of Mr. John Santee, also of Macalester, who compiled the equipment and chemical requirements which are included for each experiment in the instructor's manual. We are also most appreciative of the comments and suggestions we have received from users, both students and teachers, of the first edition, and hope that you will continue to recommend modifications in experiments that you feel would be effective.

E. J. Slowinski

W. L. Masterton

W. C. Wolsey

SAFETY IN THE LABORATORY

A chemistry laboratory can be, and should be, a safe place in which to work. Yet each year in academic and industrial laboratories accidents occur which in some cases injure seriously, or kill, chemists. Most of these accidents could have been foreseen and prevented, had the chemists involved used the proper judgment and taken proper precautions.

The experiments you will be performing have been selected at least in part because they can be done safely. Instructions in the procedures should be followed carefully and in the order given. Where it seemed likely that a very simple error could produce a dangerous set of conditions, we have noted that error specifically. Sometimes even a change of concentration of one reagent is sufficient to change the conditions of a chemical reaction so as to make it occur in a different way, perhaps at a highly accelerated rate. Do not deviate from the procedure in the manual when performing experiments unless specifically told to do so by your instructor.

One of the simplest, and most important, things you can do to avoid injury in the laboratory is to protect your eyes by routinely wearing safety glasses. If you wear prescription eyeglasses, they will serve the purpose; otherwise, use the plastic goggles that are available or buy a pair for your personal use.

The most common accident in the general chemistry laboratory occurs when a student tries to insert glass tubing, a thermometer, or a glass rod into a hole in a rubber stopper. The glass breaks because the student subjects it to excess force in the wrong direction, and sharp glass cuts the student's finger or hand, sometimes severely. Such accidents are completely unnecessary. When you are trying to put a piece of tubing into a rubber stopper, use common sense; we offer the following procedure for your consideration:

1. Make sure the hole in the stopper is only a little bit smaller than the tubing you are working with.
2. Keep both hands close to the stopper when working the tubing into the stopper. Don't hold a foot long piece of tubing by the end away from the stopper, but rather at a point an inch or two from the stopper.
3. Use a lubricant. A drop or two of glycerine, or even water, will make it much easier to work the tubing into the stopper.

Your laboratory assistant should show you the procedure to be used when you work with tubing and stoppers for the first time. If you do cut yourself go to the instructor, and he will decide on the proper treatment.

Although other kinds of accidents are less frequent, they do occasionally occur, and should be considered as possibilities. Your volatile organic liquid may ignite if you bring an open flame too close. You may spill a caustic reagent on yourself or your neighbor, or you may somehow get some chemical into your eye or mouth. A common response in such a situation is panic. All too frequently a student will, in the excitement of the incident, do something utterly irrational, such as running screaming from the room when the remedy for the accident was very close at hand. If you see an accident happen to another student, watch for signs of panic and tell the student what he should do; if it seems necessary, help him to do it. Call your instructor for assistance. Chemical spills are best handled by washing the area quickly with water from the nearest sink. More severe spills can be treated by using the showers or eye washes that may be available in your laboratory. In case of a fire, in a beaker, on the bench, on your clothing or that of another student, do not panic and run. Smother the fire with an extinguisher, a blanket, or with water, as seems most appropriate at the time. If the fire is in a piece of equipment or on the lab bench, and does not appear to require instant action, have your instructor put the fire out. In any event, learn where shower, eye wash, and the fire extinguishers are, so that you will not have to look all over if you ever need them in a hurry.

In this laboratory manual we have attempted to describe safe procedures and to employ chemicals that are safe when used properly. Many thousands of students have performed the experiments without having accidents, so you can too. However, we authors cannot be present in the laboratory when you carry out the experiments to be sure that you observe the necessary precautions. You and your laboratory supervisor must, therefore, see to it that the experiments are done properly, and assume responsibility for any accidents or injuries which may occur.

CONTENTS

SECTION SEVEN • PROPERTIES OF LIQUIDS AND SOLUTIONS

SECTION EIGHT • CHEMICAL EQUILIBRIUM

SECTION NINE • CHEMICAL KINETICS

SECTION TEN • PRECIPITATION REACTIONS

SECTION FOURTEEN • NUCLEAR CHEMISTRY

SECTION FIFTEEN • QUALITATIVE ANALYSIS

PROPERTIES OF MATTER. CHEMICAL SEPARATIONS

EXPERIMENT

1 • The Densities of Liquids and Solids

One of the fundamental properties of any sample of matter is its density, which is its mass per unit of volume. In terms of the base units of the International System, the density of water is 1000.00 kg/m³ at 4°C. In more common units it is expressed as 1.00000 g/cm³. Densities of liquids and solids range from values less than that of water to values considerably greater than that of water. Osmium metal has a density of 22.5 g/cm³ and is probably the densest material known at ordinary pressures. The densities of liquids and solids change with changes in temperature, in general decreasing slowly with increasing temperature, and slightly increasing with increasing pressure under ordinary conditions. Any change in the density of a given sample results from a change in volume, since the mass of a sample is not a function of temperature or pressure. The densities of gases can vary considerably with either pressure or temperature changes. Gas densities will be the subject of a later experiment.

In any density determination, two quantities must be determined—the mass and the volume of a given quantity of matter. The mass can easily be determined by finding the "weight" of the substance on a balance. The quantity we usually think of as "weight" is really the mass of a substance. In the process of "weighing," we find the mass, taken from a standard set of masses, that experiences the same gravitational force as that experienced by the given quantity of matter we are "weighing." The mass of a sample of liquid in a container can be found by taking the difference between the mass of the container plus the liquid and the mass of the empty container.

The volume of a liquid can easily be determined by means of a calibrated container. In the laboratory a graduated cylinder is often used for routine measurements of volume. Accurate measurement of liquid volume is made by using a pycnometer, which is simply a container having a precisely definable volume. The volume of a solid can be determined by direct measurement if the solid has a regular geometrical shape. Such is not usually the case, however, with ordinary solid samples. A convenient way to determine the volume of a solid is to measure accurately the volume of liquid displaced when an amount of the solid is immersed in the liquid. The liquid used in such an experiment should not react with or dissolve the solid and, as you may readily surmise, should have a lower density than the solid.

EXPERIMENTAL PROCEDURE

A. Mass of a Slug.   After you are shown how to operate the analytical balance in your laboratory, obtain a numbered metal slug from your instructor. Weigh it on the balance to the nearest 0.001 g. Record the mass and the number of the slug and report it to your instructor. When he has approved your weighing, go to the stockroom and obtain a glass-stoppered flask, which will serve as a pycnometer, and samples of an unknown liquid and an unknown metal.

B. Density of a Liquid.   If your flask is not clean and dry, clean it with soap and water, rinse it with a little acetone, and dry it by letting it stand for a few minutes in the air or by gently blowing compressed air into it for a few moments.

Weigh the dry flask with its stopper on the analytical balance, or the toploading balance if so directed, to the nearest milligram. Fill the flask with distilled water until the liquid level is nearly to the top of the ground surface in the neck. Put the stopper in the flask in order to drive out *all* the air and any excess water. Work the stopper gently into the flask, so that it is firmly seated in position. Wipe any water from the outside of the flask with a towel and soak up all excess water from around the top of the stopper.

Again weigh the flask, which should be completely dry on the outside and full of water, to the nearest milligram. Given the density of water at the temperature of the laboratory and the mass of water in the flask, you should be able to determine the volume of the flask very precisely. Empty the flask, dry it, and fill it with your unknown liquid. Stopper and dry the flask as you did when working with the water and then weigh the stoppered flask full of the unknown liquid, making sure its surface is dry. This measurement, used in conjunction with those you made previously, will allow you to find accurately the density of your unknown liquid.

C. Density of a Solid.   Pour your sample of liquid from the flask into its container. Rinse the flask with a small amount of acetone and dry it thoroughly. Add small chunks of the metal sample to the flask until the flask is about half full. Weigh the flask, with its stopper and the metal, to the nearest milligram.

Leaving the metal in the flask, fill the flask with water and then replace the stopper. Roll the metal around in the flask to make sure that no air remains between the metal pieces. Refill the flask if necessary, and then weigh the dry, stoppered flask full of water plus the metal sample. Properly done, the measurements you have made in this experiment will allow a calculation of the density of your metal sample that will be accurate to about 0.1 per cent.

Pour the water from the flask. Put the metal in its container. Dry the flask and return it with its stopper and your metal sample to the stockroom.

DATA AND CALCULATIONS: Densities of Liquids and Solids

Metal slug no. _____ Mass of slug _____g

Unknown liquid no. _____ Unknown solid no. _____

Density of unknown liquid

    Mass of empty flask plus stopper   _____g

    Mass of stoppered flask plus water   _____g

    Mass of stoppered flask plus liquid   _____g

    Mass of water   _____g

    Volume of flask (density of H_2O at 25°C,
    0.9970 g/cm³, at 20°C, 0.9982 g/cm³)   _____cm³

    Mass of liquid   _____g

    Density of liquid   _____g/cm³

    To how many significant figures can the liquid density
    be properly reported?   _____

Density of unknown metal

    Mass of stoppered flask plus metal   _____g

    Mass of stoppered flask plus metal plus water   _____g

    Mass of metal   _____g

    Mass of water   _____g

    Volume of water   _____cm³

    Volume of metal   _____cm³

    Density of metal   _____ g/cm³

    Would you expect the per cent error in the metal density to be higher or lower than the per cent error in the liquid density as obtained in this experiment? _____

    Why?

3

ADVANCE STUDY ASSIGNMENT: Densities of Solids and Liquids

1. In an experiment a student was asked to measure the densities of an organic liquid and a metal. He was given a small flask with a ground-glass stopper, which he found weighed 31.601 g when empty. He filled the flask with water and weighed the full, stop-pered flask, recording a mass of 60.735 g in a room at about 25° C. He dried the flask, filled it with the organic liquid, and found that the flask then weighed 56.796 g. Then he dried the flask again and added some of the metal to it; the flask plus the metal weighed 99.323 g. He filled the flask with water, leaving the metal in the flask, and found that after he had removed all the air bubbles, the flask with the metal and the water weighed 120.827 g. From the data the student obtained, calculate the volume of the flask, the density of the metal, and the density of the organic liquid.

Volume of flask   _____ cm³

Density of organic
liquid   _____ g/cm³

Density of metal _____ g/cm³

2. Give the effects of the following errors on the value of the density of the metal as ob-tained in Problem 1. Explain your reasoning in each case.
    a. The flask was not quite dry when weighed the first time.

    b. The value recorded for the mass of the flask plus the metal was 98.323 g rather than 99.323 g.

5

EXPERIMENT

2 • Resolution of Matter into Pure Substances, I. Fractional Crystallization

One of the important problems faced by chemists is that of determining the nature and state of purity of the substances with which they work. In order to perform meaningful experiments, chemists must ordinarily use essentially pure substances, which are often prepared by separation from complex mixtures.

In principle the separation of a mixture into its component substances can be accomplished by carrying the mixture through one or more physical changes, experimental operations in which the nature of the components remains unchanged. Because the physical properties of various pure substances are different, physical changes frequently allow an enrichment of one or more substances in one of the fractions that is obtained during the change. Many physical changes can be used to accomplish the resolution of a mixture, but in this experiment we will restrict our attention to one of the simpler ones in common use.

The solubilities of solid substances in different kinds of liquid solvents vary widely. Some substances are essentially insoluble in all known solvents; the materials we classify as macromolecular are typical examples. Most materials are noticeably soluble in one or more solvents. Those substances that we call salts often have very appreciable solubility in water but relatively little solubility in any other liquids. Organic compounds, whose molecules contain carbon and hydrogen atoms as their main constituents, are often soluble in organic liquids such as benzene or carbon tetrachloride.

We also often find that the solubility of a given substance in a liquid is sharply dependent on temperature. Most substances are more soluble in a given solvent at high temperatures than at low temperatures, although there are some materials whose solubility is practically temperature-independent and a few others that become less soluble as temperature increases.

By taking advantage of the differences in solubility of different substances we often find it possible to separate the components of a mixture in essentially pure form.

In this experiment you will be given a sample containing silicon carbide, potassium dichromate, and sodium chloride. Your problem will be to separate this mixture into its component parts, using water as a solvent. Silicon carbide SiC is a macromolecular substance and is insoluble in water. Potassium dichromate $K_2Cr_2O_7$ and sodium chloride $NaCl$ are water soluble ionic substances, with different solubilities at different temperatures, as indicated in Figure 2.1. Sodium chloride exhibits little change in solubility between $0°C$ and $100°C$, whereas the solubility of potassium dichromate increases about 16-fold over that temperature range. Given a water solution containing equal masses of $NaCl$ and $K_2Cr_2O_7$ it should be clear that $K_2Cr_2O_7$ would most easily crystallize in pure form from the solution at low temperatures, and that at high temperatures the crystals that would first appear in the boiling solution would be essentially pure $NaCl$. The method by which we recover pure substances by making use of solubility properties such as those cited for $NaCl$ and $K_2Cr_2O_7$ is called fractional crystallization, and this is one of the fundamental procedures used by chemists for isolating pure materials.

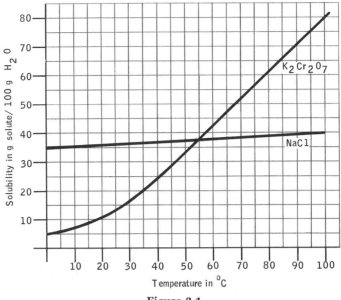

Figure 2.1

EXPERIMENTAL PROCEDURE

Obtain from the stockroom a Buchner funnel, a suction flask and a sample (about 25 g) of your unknown solid. Weigh the sample into a 250 cm³ beaker (±0.1 g). Add about 100 cm³ of distilled water, which will be enough to dissolve the soluble solids.

Support the beaker with its solution on a piece of wire gauze over an iron ring, and warm gently to about 40° C. Stir the solution to make sure that all soluble material is dissolved. Remove the insoluble silicon carbide by filtering the solution through a Buchner funnel with gentle suction. Transfer as much as you can of the solid carbide to the funnel with your rubber policeman. Transfer the orange filtrate to a clean 250 cm³ beaker. Reassemble the Buchner funnel, apply suction, and wash the SiC on the filter paper with distilled water. Continue suction for several minutes to dry the SiC. With the help of your spatula, lift the filter paper and the SiC crystals from the funnel and put the paper on the lab bench so that the crystals may dry in the air.

Heat the orange filtrate in the beaker to the boiling point and boil gently until white crystals of NaCl are visible in the liquid. The solution will have a tendency to bump, so do not heat it too strongly. Hot dichromate solution can give you a bad burn. When NaCl crystals are clearly apparent (the solution will usually appear cloudy at that point), stop heating and add 10 cm³ of distilled water to the solution. (This will be enough to dissolve the NaCl and prevent its crystallizing with the $K_2Cr_2O_7$.) Wash the crystallized solids from the walls of the beaker with your medicine dropper, using the solution in the beaker. Stir the solution with a glass rod to dissolve the solids; if necessary, you may heat the solution, but do not boil it.

Cool the solution to room temperature in a water bath, and then to about 0°C in an ice bath. Bright orange crystals of $K_2Cr_2O_7$ will precipitate. Stir the cold slurry of crystals for several minutes. Assemble the Buchner funnel; chill it by adding 100 cm³ ice-cold distilled water and, after a minute, drawing the water through with suction. *Discard* the water in the suction flask. Filter the $K_2Cr_2O_7$ slurry through the cold Buchner funnel into the empty flask. Your rubber policeman may be helpful when transferring the last of the crystals. Press the crystals dry with a clean piece of filter paper, and continue to apply suction for a minute or so. Turn off the suction and pour the filtrate, which contains most of the NaCl in the sample, into a clean 150 cm³ beaker. Heat the filtrate to boiling and boil gently.

Reassemble the Buchner funnel while you are waiting for the filtrate to boil, and, *without applying suction*, add a little ice-cold distilled water from your wash bottle to the funnel; use just enough water to cover the crystals. Let the cold liquid remain in contact with the crystals for about ten seconds, and then apply suction; the liquid removed will contain most of the NaCl impurity. Continue to apply suction for about a minute to dry the purified $K_2Cr_2O_7$ crystals. Lift the filter paper and the crystals from the funnel and put the paper on the lab bench to let the crystals dry further in the air.

Continue to boil the solution containing the NaCl until the volume of NaCl crystals (which appear as the boiling proceeds) is about equal to half the volume of liquid above the crystals. Again, wash the crystallized solids from the walls of the beaker into the solution with your medicine dropper filled with solution from the beaker. Stir to dissolve any solid $K_2Cr_2O_7$ that may be present. Reheat the solution to the boiling point.

Assemble the Buchner funnel, turn on the suction, and, using your tongs to hold the beaker, filter the hot solution through the funnel. Transfer as much of the solid NaCl crystals as possible, first by swirling the slurry as you pour it, and then by using your rubber policeman. Press the crystals flat with a piece of dry filter paper. At this point the NaCl will appear yellow because of the presence of residual dichromate solution.

Turn off the suction by disconnecting the hose from the suction flask. Add about 15 drops 6 *M* HCl drop by drop to the crystals. Wait about half a minute, and then reapply suction for a minute or so to remove the liquid, which will contain most of the yellow contaminant. Wash the crystals, with suction on, using a little acetone, which will remove the residual HCl and will readily evaporate. If this operation has been done properly your purified NaCl crystals will be nearly colorless. Remove the filter paper with the crystals from the funnel and put it aside to let the NaCl dry further.

Weigh your dry SiC and $K_2Cr_2O_7$ crystals on separate preweighed sheets of paper, using the top loading or triple beam balance. Show your samples of SiC, $K_2Cr_2O_7$, and NaCl to your laboratory supervisor for his evaluation.

DATA AND CALCULATIONS: **Resolution of Pure Substances, I.**
Fractional Crystallization

Unknown no. _____

Mass of sample and container _____g

Mass of container _____g

Mass of sample _____g

Mass of paper plus SiC _____g

Mass of paper _____g

Mass of SiC _____g

Mass of paper plus $K_2Cr_2O_7$ _____g

Mass of $K_2Cr_2O_7$ _____g

Mass of $K_2Cr_2O_7$ in original sample if you recovered 80 per cent
of it in this experiment. _____g

Per cent SiC in sample _____%

Per cent $K_2Cr_2O_7$ in sample (assuming 80 per cent recovery) _____%

Per cent NaCl in sample (by difference) _____%

ADVANCE STUDY ASSIGNMENT: **Resolution of Matter into Pure Substances, I. Fractional Crystallization**

1. From a mixture of SiC, $K_2Cr_2O_7$, and NaCl weighing 14.7 g, 3.9 g SiC and 8.5 g $K_2Cr_2O_7$ were recovered. Assuming that all of the SiC but only 80 per cent of the $K_2Cr_2O_7$ originally present were recovered and that the remainder of the sample was NaCl, what are the mass percentages of each component in the original mixture?

_____per cent SiC, _____per cent $K_2Cr_2O_7$

_____per cent NaCl

2. What solid would crystallize first from a solution containing equal amounts of $K_2Cr_2O_7$ and NaCl if the solution was evaporated at:

   a. 100°C

   b. 75°C

   c. 25°C

3. A solution containing 25 g $K_2Cr_2O_7$ and 25 g NaCl in 100 g H_2O is cooled to 0°C. How much $K_2Cr_2O_7$ will crystallize out? How much NaCl? What per cent of the $K_2Cr_2O_7$ could be recovered by this crystallization?

EXPERIMENT

3 • Resolution of Matter into Pure Substances, II. Paper Chromatography

The fact that different substances have different solubilities in a given solvent can be used in several ways to effect a separation of substances from mixtures in which they are present. We have seen in a previous experiment how fractional crystallization allows us to obtain pure substances by relatively simple procedures based on solubility properties. Another widely used resolution technique, which also depends on solubility differences, is chromatography.

In the chromatographic experiment a mixture is deposited on some solid adsorbing substance, which might consist of a strip of filter paper, a thin layer of silica gel on a piece of glass, some finely divided charcoal packed loosely in a glass tube, or even some microscopic glass beads coated very thinly with a suitable adsorbing substance and contained in a piece of copper tubing.

The components of a mixture are adsorbed on the solid to varying degrees, depending on the nature of the component, the nature of the adsorbent, and the temperature. A solvent is then caused to flow through the adsorbent solid under applied or gravitational pressure or by the capillary effect. As the solvent passes the deposited sample, the various components tend, to varying extents, to be dissolved and swept along the solid. The rate at which a component will move along the solid depends on its relative tendency to be dissolved in the solvent and adsorbed on the solid. The net effect is that, as the solvent passes slowly through the solid, the components separate from each other and move along as rather diffuse zones. With the proper choice of solvent and adsorbent, it is possible to resolve many complex mixtures by this procedure. If necessary, we can usually recover a given component by identifying the position of the zone containing the component, removing that part of the solid from the system, and eluting the desired component with a suitable good solvent.

The name given to a particular kind of chromatography depends upon the manner in which the experiment is conducted. Thus, we have column, thin-layer, paper, and vapor chromatography, all in very common use. Chromatography in its many possible variations offers the chemist one of the best methods, if not the best method, for resolving a mixture into pure substances, regardless of whether that mixture consists of a gas, a volatile liquid, or a group of nonvolatile, relatively unstable, complex organic compounds.

In this experiment we shall use paper chromatography to resolve a mixture of substances known as acid-base indicators. These materials are typically brilliant in color, with the colors depending on the acidity of the system in which they are present. A sample containing a few micrograms of the indicator is placed near one end of a strip of filter paper. That end of the paper is then immersed vertically in a solvent. As the solvent rises up the paper by capillary action it tends to carry the sample along with it, to a degree that depends on the solubility of the sample in the solvent and its tendency to adsorb on the paper. When the solvent has risen a distance of L centimetres, the solute, now spread into a somewhat diffuse zone or band, will have risen a smaller distance, say D centimetres. It is found that D/L is, for a given substance under specified conditions, a constant independent of the relative amount of that substance or other substances present. D/L is called the R_f value for that substance under the experimental conditions:

$$R_f = \frac{D}{L} = \frac{\text{distance solute moves}}{\text{distance solvent moves}}$$

The R_f value is a characteristic of the substance in a given chromatography experiment, and can be used to test for the presence of a particular substance in a mixture of substances with different R_f values.

The first part of the experiment will involve the determination of the R_f values for five common acid-base indicators. These substances have colors that will allow you to establish the positions of their bands at the conclusion of the experiment. When you have found the R_f value for each substance by studying it by itself, you will use these R_f values to analyze an unknown mixture.

EXPERIMENTAL PROCEDURE   WEAR YOUR SAFETY GLASSES WHILE PERFORMING THIS EXPERIMENT

Take a clean dry beaker and a clean dry test tube to the stockroom and obtain six paper strips and a sample of your unknown. Handle the strips carefully; whenever you need to work with them, handle them by their edges, because their surfaces can very easily be contaminated by your fingers.

Place the strips on a clean dry sheet of paper and make a pencil mark about 2 cm from one end of each strip (Figure 3.1).

Put two or three drops of the following indicators into separate, clean, dry, micro test tubes:

> Bromthymol blue   Phenolphthalein
> Alizarin yellow   Phenol red
> Bromcresol purple

For an applicator use a fine capillary tube, which will be furnished to you by your instructor. Test the application procedure by dipping the applicator into one of the colored solutions and touching it momentarily to a round test piece of filter paper. The liquid from the applicator should form a spot no larger than 0.5 cm in diameter. Test this procedure several times.

Clean the applicator by dipping it in a little acetone and blowing air through it to dry it. Dip it in one of the indicator solutions and put a 0.5 cm spot on the pencil line on one of the strips. Label the strip at point X (Figure 3.1) with the name of the indicator you applied. Clean the applicator and repeat the procedure on the other strips for each of the other indicators and the unknown. Apply the unknown three times for greater color intensity, *making sure* each time that the spot has *dried* before making the next application.

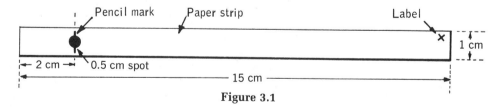

Figure 3.1

Draw 50 cm³ of eluting solvent from the supply on the reagent shelf. This solution is made by saturating normal butanol, an organic alcohol, with 1.5 M NH$_3$ solution. Pour this solution into three dry 250 cm³ Erlenmeyer flasks. These will serve as developing chambers in the experiment.

When you are sure the sample spots are dry, place two of the strips opposite each other on the side of a cork, and place the cork and the strips in the flask so that the ends of the strips, but not the sample spots, are in the solvent (see Figure 3.2). In the same way, place two strips in the second flask and two in the third.

Let solvent rise on the strips for 45 to 60 min, or until the solvent front has moved at least 9 cm above the pencil lines. Remove the strips from the beakers and put them on the sheet of paper used earlier. Draw a pencil line along the solvent front on each of the strips and let the strips dry for several minutes.

When the strips have dried (you may need to hold them in the warm air over a piece of asbestos screen held over a small Bunsen flame), hold each strip over the open mouth of a bottle of 15 M NH₃, and note the color of the band associated with each indicator when it is in this alkaline vapor. You will find that moistening the phenolphthalein strip with a damp paper towel prior to exposure to NH₃ produces a much more intense color. When you are sure that you know the position of the band to be associated with each indicator, measure the distance from the center of the band to the point where the indicator was applied. For each indicator, also measure the distance from the solvent front to the point of application. Calculate R_f values for each indicator.

On the strip containing the unknown, measure the R_f values and the colors for each band in the presence of NH₃ vapor, and identify those indicators which are in the unknown. The detection of phenolphthalein is again facilitated if the strip is moistened prior to exposure to NH₃ vapor.

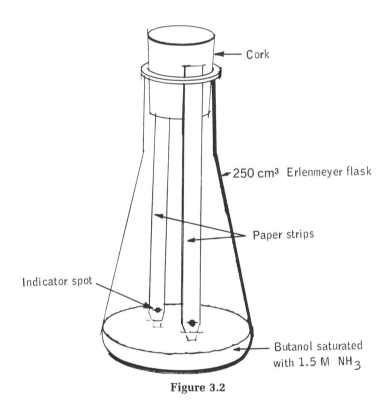

Figure 3.2

DATA AND CALCULATIONS: **Resolution of Matter into Pure Substances, III.**
Paper Chromatography

	Distance solvent moved (cm)	Distance sample moved (cm)	R_f	Color in NH_3 vapor
Bromthymol blue	_____	_____	_____	_____
Alizarin yellow	_____	_____	_____	_____
Bromcresol purple	_____	_____	_____	_____
Phenol red	_____	_____	_____	_____
Phenolphthalein	_____	_____	_____	_____
Unknown	_____	_____	_____	_____
		_____	_____	_____
		_____	_____	_____
		_____	_____	_____
		_____	_____	_____

Composition of unknown _____

Unknown no. _____

ADVANCE STUDY ASSIGNMENT: **Resolution of Matter into Pure Substances, II.**
                                            Paper Chromatography.

1. A student chromatographs a mixture, and after developing the spots with a suitable reagent he observes the following:

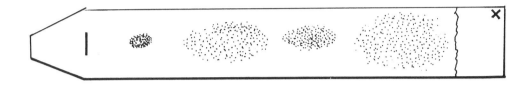

What are the R_f values?

2. Why should a pencil line be drawn to mark the spot to apply the samples in this experiment, rather than ink from a ball point or fountain pen?

3. The solvent moves the first 2 cm in about 10 min. Why shouldn't the experiment be stopped at that time instead of waiting 45 to 60 min for the solvent to move 8 to 10 cm?

4. Chromatography gets its name from the fact that, as in this experiment, the positions of the bands obtained were first identified by the colors of the resolved substances. Modern chromatographic methods are ordinarily applied to substances that are color-less in the mixtures to be resolved. Can you suggest three methods that might be used to render apparent the positions of bands containing colorless substances?

FORMULAS AND STOICHIOMETRY

EXPERIMENT

4 • Preparation of Metal Sulfides and Determination of the Gram Equivalent Mass of a Metal

An early development in chemistry that brought some ordering to the composition of compounds was the recognition that there was a definite mass ratio between any two elements in a compound. This relationship, known as the Law of Definite Proportions, was arrived at after many careful preparations and analyses of compounds.

By defining the gram equivalent mass (GEM) of an element as that amount of the element which will combine with 8.000 g of oxygen, it was possible to establish the gram equivalent masses of many elements. Since some elements form more than one oxide, these elements had two, or even three, gram equivalent masses. The usefulness of gram equivalent masses lay in the fact that they could be applied to compounds that did not contain oxygen. Knowing, for example, that the gram equivalent mass of magnesium was 12.16 and sulfur 16.03 g, one could predict that these two elements would form a compound in which the mass ratio of magnesium to sulfur was 12.16:16.03, the general rule being that one GEM of an element combines with one GEM of another. The drawback to this approach was that many elements had more than one GEM, so that prediction of mass ratios of elements in compounds could produce several possible values rather than one. The development of the idea of atomic masses and chemical formulas eliminated the need for gram equivalent masses, but the early chemists used them extensively, since they did systematize to some extent mass relationships among elements.

The gram equivalent mass of an element, unlike its atomic mass, can be found by chemical analysis of one of its binary compounds. In this experiment you will determine the gram equivalent mass of a metal by preparing its sulfide, using a known mass of the metal and an excess of sulfur. The excess sulfur will be burned off as SO_2, and the metal will combine completely with sulfur. The amount of sulfur that is present in the sulfide which is formed can be determined by taking the difference between the mass of the sulfide product and that of the metal used. Since the gram equivalent mass of sulfur in the sulfide is known to be 16.03 g, the gram equivalent mass of the metal, that mass which combines with 16.03 g of sulfur, can be found by a simple calculation.

(Ideally, this method should give a very precise value for the gram equivalent mass of the metal in the metal sulfide. Unfortunately, some metal sulfides, particularly those of the transition elements, are nonstoichiometric compounds in which the number of atoms of metal that will combine with one atom of sulfur may vary slightly, depending upon the conditions under which the two elements combine. The experimental ratio of atoms will, however, generally be very close to a whole number in these compounds.)

EXPERIMENTAL PROCEDURE   WEAR YOUR SAFETY GLASSES WHILE PERFORMING THIS EXPERIMENT

Obtain a sample of an unknown metal from the stockroom. Clean a porcelain crucible by adding a little 6 M nitric acid and heating gently. Rinse the crucible with water and dry it. Any stain on the porcelain that does not come off with the nitric acid treatment will not interfere in this experiment. Place the crucible on a clay triangle supported on a ring stand. Heat the crucible with a Bunsen burner, gently at first, then to a dull red color for 5 min. Allow the crucible to cool to room temperature, and then weigh the crucible and cover to the nearest 0.001 g. Always use your crucible tongs to handle the crucible.

Place the sample of your unknown metal in the crucible. If the metal is a powder, make sure that none of it sticks to the sides. If the metal is a wire, coil it so that it all fits well down into the crucible. Weigh the covered crucible containing the sample. Cover the metal sample with powdered sulfur to a depth of about 5 mm. In the hood, heat the covered crucible with a small flame until the characteristic blue color of burning sulfur is no longer visible around the edge of the crucible cover. Heat the crucible with a strong flame for 10 min more. Allow the crucible to cool to room temperature and reweigh the covered crucible and contents.

All of the metal may not have reacted with the sulfur the first time, and it will be necessary to check for completeness of reaction. Cover the contents of the crucible with another layer of sulfur and heat as before. Cool and weigh the crucible and contents a second time. The masses should agree within 5 mg. If the masses differ by more than this amount, repeat the sulfur ignition a third time.

Add a few drops of 6 M HCl to the contents of the crucible and note the odor of the gas evolved. What is this gas?

DATA: Preparation of Metal Sulfide

Mass of crucible and cover   _____g

Mass of crucible, cover, and metal   _____g

Mass of crucible, cover, and metal sulfide (first heating)   _____g

Mass of crucible, cover, and metal sulfide (second heating)   _____g

Mass of crucible, cover, and metal sulfide (third heating if needed)   _____g

Mass of metal   _____g

Mass of metal sulfide   _____g

Mass of combined sulfur   _____g

Hence, _____g metal $\simeq$ _____g sulfur

_____g metal $\simeq$ 16.03 g sulfur

GEM metal = _____g

Unknown no. _____

ADVANCE STUDY ASSIGNMENT: Preparation of a Metal Sulfide

1. A metal sample weighing 1.605 g was found to produce a metal sulfide weighing 1.861 g. Calculate the gram equivalent mass of the metal.

_____g

2. In a reaction with sulfur a sample of thallium has a gram equivalent mass of 204 g. What is the formula of the thallium sulfide? Atomic masses: S = 32.06, Tl = 204.4

_____

3. State whether each of the following errors in procedure would result in a value for the gram equivalent mass greater or less than the true value, or would have no effect. Support your answer with a reason in each case.

 a. The empty crucible was not dry when it was weighed.

 b. Some of the sulfur was spilled before the sample was ignited.

 c. Some of the metal did not undergo any reaction.

 d. The sample of sulfur used contained an inert impurity.

 e. Part of the metal sample was converted to the oxide and not to the sulfide.

EXPERIMENT

5 • Law of Multiple Proportions

In the previous experiment the gram equivalent mass of a metal was discussed and a simple procedure was presented for its determination. You should read the discussion in that experiment before proceeding further.

Many elements form more than one compound with oxygen or other nonmetals. These elements clearly must have more than one gram equivalent mass. The Law of Multiple Proportions states that in such cases the different gram equivalent masses of an element are in a ratio of simple whole numbers (2:1, 3:2, etc.). This experiment* is designed to illustrate the validity of the Law of Multiple Proportions and to give you some experience with simple reactions and laboratory procedures.

Copper and bromine form more than one compound with each other. A copper bromide (compound A) dissociates when heated, liberating bromine and forming another copper bromide (compound B). This second bromide is readily converted to an oxide (compound C) by treatment with nitric acid and subsequent heating. This oxide of copper can be easily reduced to copper metal with hydrogen. In this experiment you will carry a known mass of compound A through these reactions, measuring the related masses of compounds B and C. From these data, you will be able to calculate gram equivalent masses for copper and bromine. Knowing the atomic masses of copper and bromine, you can then establish the formulas of compounds A, B, and C.

EXPERIMENTAL PROCEDURE   WEAR YOUR SAFETY GLASSES WHILE PERFORMING THIS EXPERIMENT

Place about one gram of the copper bromide (A) in a large, accurately weighed test tube. Weigh the test tube and its contents to the nearest 0.001 g. Clamp the test tube so that it is slightly inclined, and incorporate it into an apparatus like that shown in Figure 5.1. The bent glass tube should extend about an inch past the end of the rubber stopper and should dip to within about 2 cm of the surface of the water in the 500 cm^3 Florence flask. Do not let the tube dip into the water; you do not want water to back up into the test tube during the experiment.

Heat the bromide sample, first gently and then rather strongly with the burner flame, until bromine, Br_2, is no longer evolved. (If the tube is heated to redness, the initial decomposition product may be further decomposed to copper, so use discretion when heating.) Carefully heat the upper end of the test tube, if necessary, to drive any condensed bromine out of the test tube. Bromine is very reactive and very poisonous. Do not inhale the bromine vapor or touch the liquid. Allow the test tube to cool and weigh it to find the amount of the second copper bromide (B) formed.

Add 2 cm^3 of 15 M HNO_3 (carefully!) to the test tube and reassemble the apparatus as before. Heat the tube gently until the black substance formed, copper oxide (C), is completely dry. Weigh the test tube and its contents again to find the amount of copper oxide formed from the decomposition of the copper nitrate produced by the reaction of nitric acid with copper bromide.

In the next part of the experiment you will reduce the copper oxide (C) to copper with hydrogen gas. Set up the hydrogen generator described in Experiment 6, Figure 6.1, as directed in that experiment. Have your instructor check your apparatus before proceeding further.

---

*Similar to an experiment described by J. C. Bailar, J. Chem. Educ. 6, 1759 (1929).

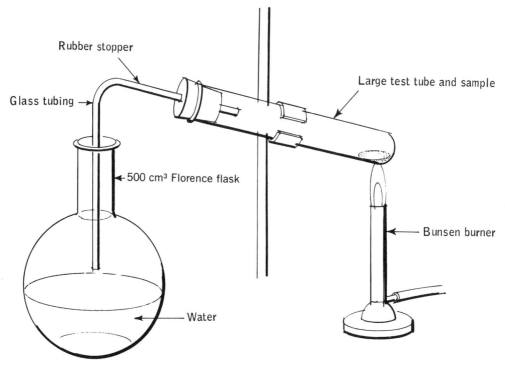

Rubber stopper

Glass tubing →

←500 cm³ Florence flask

Large test tube and sample

←Bunsen burner

Water

Figure 5.1

Follow the procedure given in Experiment 6 for the generation of hydrogen gas and its use in the reduction of metal oxides. Be very careful when igniting the hydrogen jet, and follow the instructions exactly. Try to keep the jet burning throughout the reduction and to keep the flame at a constant size by the judicious addition of acid.

When the copper oxide has been completely reduced, let the test tube cool while hydrogen is still passing through it. Weigh the test tube and its contents.

This experiment ordinarily requires two laboratory sessions. You may conveniently stop after preparing and weighing the copper oxide, or you may proceed to construct the hydrogen generator if there is time. During the second session you can complete the reduction of the oxide and make the calculations. Be sure to save your sample of copper oxide; stopper it before putting it in your locker.

DATA: Law of Multiple Proportions

Mass of empty test tube _____g

Mass of test tube plus copper bromide (A) _____g

Mass of test tube plus copper bromide (B) _____g

Mass of test tube plus copper oxide (C) _____g

Mass of test tube plus copper _____g

Mass of copper bromide (A) _____g

Mass of copper bromide (B) _____g

Mass of copper oxide (C) _____g

Mass of copper _____g

CALCULATIONS

A. Using the data you obtained and the definition of gram equivalent mass, calculate the GEM of copper in the oxide (C).

GEM Cu _____g

B. Assuming that one GEM Cu combines with one GEM Br in any copper bromide compound, calculate the GEM Br in copper bromide (A) and copper bromide (B).

GEM Br in (A) _____g

GEM Br in (B) _____g

C. How do these results illustrate the Law of Multiple Proportions?

Continued on following page   **31**

D. Given the atomic masses for copper, bromine, and oxygen, find the chemical formulas for compounds A, B, and C.

Copper bromide (A) _____

Copper bromide (B) _____

Copper oxide (C) _____

E. Write balanced chemical equations for the reactions that occurred when

1. Compound A was heated.

2. Compound C was heated in a stream of hydrogen.

Note: The calculations in this experiment are based on a constant gram equivalent mass for copper in its compounds. This leads to a set of GEM's for other elements which differ to some extent from those obtained using the usual assumption that the gram equivalent mass of oxygen is constant. The laws of chemistry are valid on either basis, and in this experiment it is certainly most convenient to assume, as we do, that the GEM of copper remains fixed.

ADVANCE STUDY ASSIGNMENT: Law of Multiple Proportions

1.   Tin forms two iodides. Iodide A contains 68.1 per cent iodine by mass and iodide B contains 81.0 per cent iodine. Tin also forms an oxide which contains 21.3 per cent oxygen. On the basis of these data,

    a. Find the gram equivalent mass of tin in its oxide.

                                                               _____g

    b. Based on the mass of tin calculated in (a), find two possible gram equivalent masses of iodine.

                                                              _____g

                                                                _____g

    c. Show how, if at all, the results of (a) and (b) relate to the Law of Multiple Proportions.

2.   In the experiment to be performed, a student did not completely reduce the copper oxide (C) to copper. How would this affect the values he obtained for the gram equivalent masses of bromine? Would the data still support the Law of Multiple Proportions?

EXPERIMENT
6 • Mass Analysis of Metal Oxides

The gram equivalent mass of an element is most simply defined as the mass of that element which will combine with 8.000 g of oxygen. Gram equivalent masses of metals can in principle be determined by starting with a known mass of metal and determining the amount of oxide produced by reaction of the metal with oxygen. In this experiment you will find the gram equivalent mass of the metal by decomposing its oxide to the free metal. The gram equivalent mass of the metal can then be calculated from the mass of the oxide sample and the amount of metal produced.[*]

A few metal oxides, such as those of silver and mercury, can be decomposed to the free metal by heat alone. All others must react with some substance, called a reducing agent, that has a stronger affinity for oxygen than the metal in the metal oxide. In preparing metals from their oxide ores, carbon in the form of coke is used whenever possible, since it is relatively inexpensive. When reduction with carbon is not feasible, either because the reaction simply will not proceed or because undesirable carbides are formed, a more active metal such as aluminum is sometimes used. The metals with the most stable oxides, such as sodium and aluminum, are typically won from their ores by electrolysis.

In the laboratory, hydrogen gas is a very convenient reducing agent. It can be easily generated by the reaction of zinc metal with sulfuric acid:

$$Zn(s) + 2\ H^+(aq) \rightarrow H_2(g) + Zn^{2+}(aq)$$

At high temperatures hydrogen will reduce many metal oxides according to the following equation:

$$M_xO_y(s) + y\ H_2(g) \rightarrow x\ M(s) + y\ H_2O(g)$$

The experiment you will perform today involves this reaction. You will carry out the experiment by heating an unknown metal oxide in a stream of hydrogen gas until the oxide is completely reduced to the metal and the water vapor produced has been removed in the gas stream.

EXPERIMENTAL PROCEDURE

WEAR YOUR SAFETY GLASSES WHILE PERFORMING THIS EXPERIMENT

Obtain from the stockroom a dropping funnel or thistle tube, a drying tube, and a sample of an unknown metal oxide.

Assemble the apparatus (Fig. 6.1) consisting of a hydrogen generator and a sample tube. Support the generator and sample tube on separate ringstands. The dropping funnel should nearly touch the bottom of the 250 cm³ Erlenmeyer flask. The jet issuing from the sample tube can be made from a piece of glass tubing by drawing out to a capillary and cutting back to a diameter of about 3 mm. The gas delivery tube from the drying tube should reach to about 5 cm from the bottom of the large test tube. Use proper caution and procedures when inserting glass tubing into the rubber stoppers. The drying tube should be filled with fresh $CaCl_2$.

---

[*]See W. L. Masterton and J. J. Demo, J. Chem. Educ. 35, 242–244, (1958).

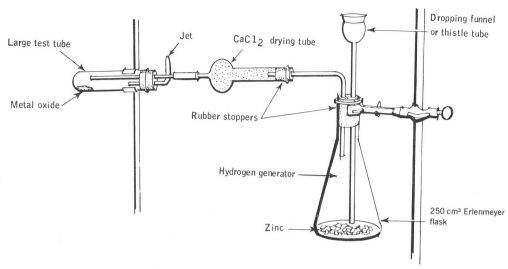

Figure 6.1

Weigh the empty test tube to 0.001 g on the analytical balance. If you use a two-pan balance, suspend the tube from the hook holding the balance pan. Place about one gram of your unknown metal oxide in the test tube and weigh it accurately.

Connect the test tube containing the sample to the generator, being careful not to let the gas delivery tube touch the sample. Add about 20 g of mossy zinc to the flask. Add a few drops of $CuSO_4$ solution and enough water to barely cover the zinc. The end of the dropping funnel or thistle tube should be under the water. Check with your instructor for approval of your apparatus before proceeding any further.

Add about 30 cm^3 of 6 M H_2SO_4 through the dropping funnel. Hydrogen gas should begin to evolve and will soon flush the air from the whole apparatus. Although a hydrogen-air explosion is very unlikely, you should wrap the hydrogen generator in a towel as a safety precaution. Let the generator run for a few minutes, and then collect a sample of the gas escaping from the jet with a small inverted test tube. Ignite the hydrogen in the test tube by holding it close to the flame from your Bunsen burner. Then try to light the jet with the burning hydrogen in the small test tube. If the hydrogen coming from the jet is pure, the flame will last long enough for you to do this. *Never* light the jet of hydrogen with a match or burner flame, since this can result in an explosion.

Once the jet is burning, heat the sample of the metal oxide with your burner flame. As the heating proceeds and the oxide is reduced, the hydrogen flame may decrease in size. Add sulfuric acid as necessary to keep the size of the flame as constant as possible. If drops of water condensing in the glass tube in the generator or in the jet reduce the flow of hydrogen, the flame may sputter or even go out. If this happens, call your instructor and ask him for advice.

Continue to heat the sample with as hot a flame as possible for about 15 min. After you stop heating, let hydrogen pass through the apparatus until the sample is at room temperature. Warm samples may reoxidize if exposed to air. Weigh the tube and contents.

If time permits (at least 45 min will be required), check for completeness of reduction of the metal oxide by heating the sample again in a stream of hydrogen, following all of the precautions taken previously. The masses of metal as calculated after the first and second heatings should agree within 0.002 g.

DATA: Mass Analysis of Metal Oxides

Mass of empty test tube   _____ g

Mass of test tube and metal oxide   _____ g

Mass of test tube and metal (first heating)   _____ g

Mass of test tube and metal (second heating)   _____ g

CALCULATIONS AND RESULTS

Mass of metal oxide   _____ g

Mass of metal contained in sample   _____ g

Mass of oxygen contained in sample   _____ g

_____ g metal $\simeq$ _____ g oxygen

_____ g metal $\simeq$ 8.000 g oxygen

GEM metal = _____ g   Per cent metal in oxide _____

Unknown no. _____

ADVANCE STUDY ASSIGNMENT: **Mass Analysis of Metal Oxides**

1.   A sample of a certain metal oxide weighing 3.728 g yields 3.344 g of metal on reaction with hydrogen. Calculate the gram equivalent mass of the metal. Can you suggest what this metal might be? The formula of the oxide? Hint: The atomic mass of the metal is greater than 150.

2.   What effect would the following errors have on the value obtained for the GEM of the metal? Explain your reasoning in each case.

    a. The sample contained an inert impurity.

    b. The oxide sample was not quite dry when weighed initially.

    c. Some water condensed on the upper end of the test tube and was weighed along with the metal.

3.   Can you suggest a reason why using the burning hydrogen to light the jet is much safer than using a match?

EXPERIMENT

7 • Water of Hydration

Most solid chemical compounds will contain some water if they have been exposed to the atmosphere for any length of time. In most cases the water is present in quite small amounts and is merely adsorbed on the surface of the crystals. This adsorbed water can usually be removed by gentle heating. Other solid compounds will contain larger amounts of water that are bound to the compound more strongly. These compounds are usually ionic salts. The water present in these salts, called water of hydration, is generally bound to the cations in the compound.

A few hydrated compounds lose water spontaneously to the atmosphere upon standing. Such compounds are called efflorescent. More generally, hydrated compounds may be dehydrated by heating. As the temperature is increased, the vapor pressure of water above the solid hydrate will increase until it exceeds the partial pressure of water in the atmosphere above it. At that temperature, dehydration occurs. As the water of hydration is lost from a hydrated compound, the compound may go through several color changes, which will correspond to the colors of the various hydrates formed by the salt. Thus $CoCl_2 \cdot 6H_2O$ is red, $CoCl_2 \cdot 2H_2O$ is violet, and $CoCl_2$ is blue.

Some anhydrous ionic compounds will absorb water from the atmosphere so strongly that they can be used to dry liquids or gases. These substances, called desiccants, are referred to as hygroscopic substances. A few ionic compounds will take up so much water from the atmosphere that they may eventually dissolve in their water of hydration; such substances are deliquescent.

In this experiment you will study the kinds of compounds that form hydrates. You will also determine the percentage of hydrate water lost by an unknown compound upon heating. You may be able to calculate the number of moles of water of hydration per mole of hydrated compound from the amount of water lost and the formula mass of the anhydrous compound.

EXPERIMENTAL PROCEDURE

WEAR YOUR SAFETY GLASSES WHILE
PERFORMING THIS EXPERIMENT

A. Identification of Hydrates. Place about 0.5 g of each of the compounds listed below in small dry test tubes. Observe carefully the behavior of each compound when you heat it gently with a burner flame. If droplets of water condense on the cool upper walls of the test tube, this is evidence that the compound may be a hydrate. Note the color and nature of the residue. See if the residue dissolves in water. If the residue is water soluble, then the original compound was probably a true hydrate. If the residue is not water soluble, then the compound is one that forms water as one of its products upon decomposition.

Sodium sulfate	Sugar
Potassium chloride	Potassium dichromate
Sodium tetraborate (borax)	Barium chloride

B. Reversibility of Hydration. Gently heat a few fine crystals of hydrated copper (II) sulfate $CuSO_4 \cdot 5H_2O$ in an evaporating dish until the color changes appear to be complete. Let a portion of the residue stand on a watch glass until the end of the laboratory period. Note any changes in color upon standing. Dissolve the rest of the residue in the evaporating dish in a few drops of water. Heat the resulting solution to boiling (CAUTION!) and allow to cool. Observe any color changes.

C. Deliquescence and Efflorescence.   Place a few crystals of each of the compounds listed here on a watch glass and allow them to stand until the end of the laboratory period. Observe them occasionally and note any changes that occur.

$Na_2CO_3 \cdot 10H_2O$ (washing soda)	$MgSO_4 \cdot 7H_2O$ (Epsom salt)
$Na_2SO_4 \cdot 10H_2O$	$KAl(SO_4)_2 \cdot 12H_2O$ (alum)
$CaCl_2$	

D. Decomposition of $AlCl_3 \cdot 6H_2O$.   Place a few crystals of $AlCl_3 \cdot 6H_2O$ in an evaporating dish. Heat gently and observe the color and odor of any vapor given off. Test the vapor with moist litmus paper to check for its acidity or basicity. If the vapor is acidic, the litmus will turn from blue to red; if it is basic, the litmus will turn from red to blue.

E. Percent Water in a Hydrate.   Clean a porcelain crucible and its cover with 6 M HNO_3. Any stain that is not removed with this treatment will not interfere in this experiment. Rinse and dry the crucible and support it on a clay triangle. Heat the crucible with a flame, gently at first, then to redness. Allow it to cool to room temperature and weigh it on the analytical balance. Handle the crucible only with your crucible tongs.

Place about one gram of your unknown solid hydrate in the crucible and again weigh the crucible on the analytical balance. Place the crucible on the triangle and put the cover on it in an off-centered position so that the water vapor can escape. Heat the crucible gently at first and then to redness for about 10 min. Cover the crucible and allow it to cool to room temperature. Weigh the cooled crucible and its contents.

If there is time, repeat the heating and weighing procedure to be sure that all of the salt is completely dehydrated. You may assume that all of the water has been removed when two successive heatings and weighings give no significant (~2 or 3 mg) change in weight.

Examine the solid residue and see if it is water soluble. Calculate the percent of water in the hydrate.

DATA AND OBSERVATIONS: Water of Hydration

A. Identification of Hydrates

	H$_2$O appears	Color of residue	Water soluble	Hydrate
Sodium sulfate	_____	_____	_____	_____
Potassium chloride	_____	_____	_____	_____
Sodium tetraborate	_____	_____	_____	_____
Sugar	_____	_____	_____	_____
Potassium dichromate	_____	_____	_____	_____
Barium chloride	_____	_____	_____	_____

B. Reversibility of Hydration

Summarize your observations on $CuSO_4 \cdot 5H_2O$:

Is the dehydration and hydration of $CuSO_4$ reversible?

C. Deliquescence and Efflorescence

	Observation	Conclusion
$Na_2CO_3 \cdot 10H_2O$	_____	_____
$Na_2SO_4 \cdot 10H_2O$	_____	_____
$CaCl_2$	_____	_____
$MgSO_4 \cdot 7H_2O$	_____	_____
$KAl(SO_4)_2 \cdot 12H_2O$	_____	_____

Continued on following page   **43**

D. Decomposition of $AlCl_3 \cdot 6H_2O$

Observations:

What substance, in addition to H_2O, would you suggest is evolved when $AlCl_3 \cdot 6H_2O$ is heated?

E. Percent Water in a Hydrate

Mass of crucible and cover   _____ g

Mass of crucible, cover, and solid hydrate   _____ g

Mass of crucible, cover, and residue (first heating)   _____ g

Mass of crucible, cover, and residue (second heating)   _____ g

CALCULATIONS AND RESULTS

Mass of solid hydrate   _____ g

Mass of residue   _____ g

Mass of H_2O lost   _____ g

Percentage H_2O in the unknown hydrate   _____ %

Formula mass of anhydrous salt (if furnished)   _____

Number of moles of water per mole of unknown hydrate   _____

Unknown no.   _____

ADVANCE STUDY ASSIGNMENT: Water of Hydration

1. A solid hydrate weighing 2.691 g was heated to drive off the water. A solid anhydrous residue remained, which weighed 2.259 g. Calculate the per cent water in the hydrate. If the anhydrous residue has a formula mass of 282, how many moles of water are present in one mole of the hydrate?

_____ % H_2O

_____ moles H_2O

2. Some hydrated metal chlorides will decompose upon heating, evolving hydrogen chloride along with water. How could one determine if this was happening to a sample of a particular metal chloride hydrate?

3. An inexpensive hygrometer often used by cello players in Minnesota is based upon the color changes of cobalt(II) chloride as the relative humidity of the atmosphere changes. When the air is dry the color is blue; when the air is moist the color is pink. Suggest an explanation of these color changes. Suggest why Minnesota cello players use hygrometers.

CALORIMETRY

EXPERIMENT
8 • Heat Effects and Calorimetry

Heat is a form of energy, sometimes called thermal energy, which can pass spontaneously from an object at a high temperature to an object at a lower temperature. If the two objects are in contact they will, given sufficient time, both reach the same temperature.

Heat flow is ordinarily measured in a device called a calorimeter. A calorimeter is simply a container with insulating walls, made so that essentially no heat is exchanged between the contents of the calorimeter and the surroundings. Within the calorimeter chemical reactions may occur or heat may pass from one part of the contents to another, but no heat flows into or out of the calorimeter from or to the surroundings.

A. Specific Heat.   When heat flows into a substance the temperature of that substance will increase. The quantity of heat Q required to cause a temperature rise Δt of any substance is proportional to the mass m of the substance and the temperature change, as shown in Equation (1). The proportionality constant C is called the specific heat of that substance.

$$Q = Cm\Delta t \qquad\qquad (1)$$

The specific heat can be considered to be the amount of heat required to raise the temperature of one gram of the substance by one degree Celsius. Amounts of heat are usually measured in joules. Specific heat will therefore have the dimensions J/g·°C. The specific heat of a substance will change slightly with temperature, but for most purposes, it can be assumed to be a constant that is independent of temperature. Water has the highest specific heat (4.18 J/g·°C) of any ordinary substance.

The specific heat of a metal can readily be measured in a calorimeter. A weighed amount of metal is heated to some known temperature and is then quickly poured into a calorimeter that contains a measured amount of water at a known temperature. Heat flows from the metal to the water and the two equilibrate at some temperature between the initial temperatures of the metal and the water.

Assuming that no heat is lost from the calorimeter to the surroundings and that a negligible amount of heat is absorbed by the calorimeter walls, the amount of heat that flows from the metal as it cools is equal to the amount of heat absorbed by the water:

$$\frac{\text{heat given off}}{\text{by the metal}} = Q = C_m m_m \mid \Delta t_m \mid = \frac{\text{heat absorbed}}{\text{by the water}} = C_{H_2O} m_{H_2O} \mid \Delta t_{H_2O} \mid \qquad (2)$$

If we measure the initial and final temperatures of the water and the metal and the masses of the water and metal used, we can use Equation (2) to find the specific heat C_m of the metal. In the first part of this experiment, you will measure the specific heat of an unknown metal by the method we have outlined.

The specific heat of a metal is related in a simple way to its atomic mass. Dulong and Petit discovered many years ago that about 25 J were required to raise the temperature of one gram atomic mass of many metals by one degree Celsius. This relation, shown in Equation (3), is known as the Law of Dulong and Petit:

$$25 \text{ J/°C} \cong C_m \times GAM_m \qquad (3)$$

Once the specific heat of a metal is known, its approximate atomic mass can be calculated from Equation (3). The Law of Dulong and Petit was one of the few rules available to guide the early chemists in their studies of atomic masses.

B. Heat of Solution.   When a solid substance and a liquid, both at the same initial temperature, are mixed and the solid dissolves in the liquid, the temperature of the solution formed is typically different from that of the initial system. The amount of heat which must flow into the solution to return it to the initial temperature of the system is called the heat of solution, $\Delta H_{solution}$, for the process. If, for example, the solution is colder than the initial system and 250 J must be furnished to the solution to return it to the original temperature, then $\Delta H_{solution}$ equals 250 J. Since heat must flow *into* the solution, the reaction is said to be *endo*thermic, and $\Delta H_{solution}$ is *positive*. If, on the other hand, the solution is warmer than the system was originally, and 320 J have to be removed from the solution to bring it to the initial temperature, then $\Delta H_{solution}$ equals −320 J. Since heat flows *from* the solution, the reaction is *exo*thermic, and $\Delta H_{solution}$ is *negative*. The heat flow for the solution of a mole of solute is called the molar heat of solution; this is the quantity that would be found in the literature for the solution reaction.

The heat of solution of a solid compound can be easily measured in a calorimetric experiment. The temperature change of the solvent water is measured and the quantity of heat that must be evolved or absorbed in returning the solution to its initial temperature is calculated from the known specific heats and the masses of the water and the solute. One can then calculate the molar heat of solution by multiplying the number of joules of heat absorbed or evolved per gram of solute by the formula mass of the solute.

EXPERIMENTAL PROCEDURE

A. Specific Heat.   From the stockroom obtain a calorimeter, a sensitive thermometer, a sample of metal in a large test tube, and a sample of unknown solid. The thermometer is very expensive, so be careful when handling it.

The calorimeter consists of two nested expanded polystyrene coffee cups fitted with a styrofoam cover. There are two holes in the cover for a thermometer and a glass stirring rod that has a loop bent on one end. Assemble the experimental setup as shown in Figure 8.1.

Weigh your sample of unknown metal in the large test tube to the nearest 0.1 g on the top loading or triple beam balance. Pour the metal into a dry container and weigh the empty test tube. Replace the metal in the test tube and put the test tube in a beaker of water. The beaker should contain enough water so that the top of the metal is below the

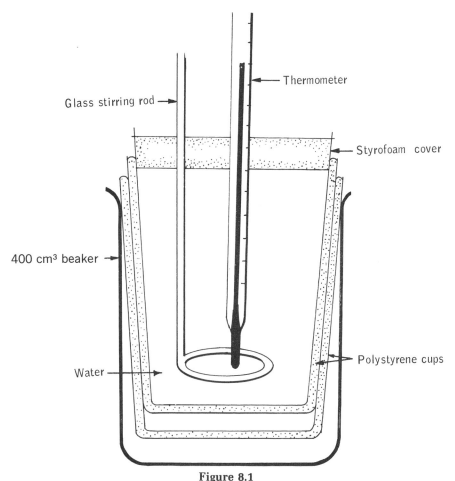

Figure 8.1

surface of the water. Heat the water to boiling and allow it to boil for a few minutes to ensure that the metal attains the temperature of the boiling water.

While the water is boiling, weigh the calorimeter to 0.1 g. Place about 40 cm³ of water in the calorimeter and weigh again. Insert the stirrer and thermometer into the cover and put it on the calorimeter. The thermometer bulb should be completely under the water.

Measure the temperature of the water in the calorimeter to 0.1°C. Take the test tube out of the beaker of boiling water and quickly pour the metal into the water in the calorimeter. Be careful that no water adhering to the outside of the test tube runs into the calorimeter when you are pouring the metal. Replace the calorimeter cover and agitate the water as best you can with the glass stirrer. Record to 0.1°C the maximum temperature reached by the water. Repeat the experiment, using about 50 cm³ of water in the calorimeter. Be sure to dry your metal before reusing it; this can be done by heating the metal briefly in the test tube in boiling water and then pouring the metal onto a paper towel to drain. You can dry the hot test tube with a little compressed air.

The metal used in this part of the experiment is to be returned to the stockroom in the test tube in which you obtained it.

B. Heat of Solution. Place about 50 cm³ of distilled water in the calorimeter and weigh as in the previous procedure. Measure the temperature of the water to 0.1°C. The temperature should be within a degree or two of room temperature. In a small beaker weigh out about 5 g of the solid compound assigned to you. Make the weighing of the beaker and of the beaker plus solid to 0.1 g. Add the compound to the calorimeter. Stirring continuously and occasionally swirling the calorimeter, determine to 0.1°C the maximum or minimum temperature reached as the solid dissolves. Check to make sure that all the solid dissolved. A temperature change of at least five degrees should be obtained in this experiment. If necessary, repeat the experiment, increasing the amount of solid used.

DATA AND CALCULATIONS: Calorimetry

A. Specific Heat	Trial 1	Trial 2

A. Specific Heat   Trial 1   Trial 2

Mass of test tube plus metal   _____g $\longrightarrow$ _____g

Mass of test tube   _____g $\longrightarrow$ _____g

Mass of calorimeter   _____g $\longrightarrow$ _____g

Mass of calorimeter and water   _____g   _____g

Mass of water   _____g   _____g

Mass of metal   _____g $\longrightarrow$ _____g

Initial temperature of water in
calorimeter   _____°C   _____°C

Initial temperature of metal (assume
100°C unless directed to do otherwise) _____°C $\longrightarrow$ _____°C

Equilibrium temperature of metal
and water in calorimeter   _____°C   _____°C

Amount of heat gained by the water _____J   _____J
(C_{H_2O} = 4.18 J/g·°C)

Amount of heat lost by the metal   _____J   _____J

Specific heat of the metal   _____J/g°C   _____J/g°C

Approximate atomic mass of metal   _____   _____

Unknown no.   _____

B. Heat of Solution

Mass of calorimeter plus water   _____g

Mass of beaker   _____g

Mass of beaker plus solid   _____g

Continued on following page   **51**

Mass of water (M_w)   _____ g

Mass of solid (M_s)   _____ g

Original temperature (T_1)   _____ °C

Final temperature (T_2)   _____ °C

$Q_w = M_w(T_1 - T_2) \ (4.18 \ \text{J/g} \cdot \text{°C})$ _____ J

$Q_s = M_s(T_1 - T_2) \ (1 \ \text{J/g} \cdot \text{°C})$*   _____ J

$Q = Q_w + Q_s =$ heat flow into solution   _____ J

The quantity you have just calculated is approximately equal to the heat of solution of your sample. Note that if $T_1 > T_2$, the reaction is endothermic, heat has to be absorbed to return the system to its original temperature, and Q has a positive sign. If $T_1 < T_2$, the reaction is exothermic and Q is a negative quantity.

Calculate the heat of solution per gram of solid.

ΔH per gram = _____ J

Solid Unknown No. _____

(Optional)

Formula of substance used (if furnished by instructor)   _____

Formula mass of compound   _____ g

Heat of solution per mole of compound, ΔH_{molar} _____ J

---

*The specific heat will differ somewhat depending upon the nature of the solid, but the value used here, 1 J/g · °C, is close enough for our purposes in this experiment.

ADVANCE STUDY ASSIGNMENT: Heat Effects and Calorimetry

1.   A metal sample weighing 44.6 g and at a temperature of 100°C was placed in 35.2 g of water contained in a calorimeter at 24.7°C. At equilibrium the temperature of the water and metal was 34.2°C. What is the specific heat of the metal? What is its approximate atomic mass? (C_{H_2O} = 4.18 J/g · °C)

Specific heat _____ J/g°C

Atomic mass _____

2.   When 4.0 g of LiI were dissolved in 48 cm³ of water in a calorimeter at 24.8°C, the temperature of the solution rose to 33.7°C. Following the procedure described on p. 52, calculate the heat of solution per gram of LiI and its molar heat of solution.

ΔH per gram_____

ΔH per mole_____

3.   In the experiment we assume that the calorimeter is not only a good heat insulator but also that it absorbs only a very small amount of heat from its contents. Although these are good approximations for a calorimeter made of expanded polystyrene, some heat will actually be lost to the surroundings, and some will be absorbed by the calorimeter walls and the thermometer if the contents of the calorimeter are above room temperature. Would you expect that these effects would result in specific heat values that are larger or smaller than the true values? Why?

PROPERTIES OF GASES

EXPERIMENT

9 • Analysis of an Aluminum-Zinc Alloy*

Some of the more active metals will react readily with solutions of strong acids, producing hydrogen gas and a solution of a salt of the metal. In a previous experiment you generated hydrogen by the action of sulfuric acid on metallic zinc:

$$Zn(s) + 2H^+(aq) \rightarrow H_2(g) + Zn^{2+}(aq) \qquad (1)$$

From this equation it is clear that one mole of zinc produces one mole of hydrogen gas in this reaction. If the hydrogen were collected under known conditions, it would be possible to calculate the mass of zinc in a pure sample by measuring the amount of hydrogen it produced on reaction with acid.

Since aluminum reacts spontaneously with strong acids in a manner similar to that shown by zinc,

$$2\,Al(s) + 6\,H^+(aq) \rightarrow 2\,Al^{3+}(aq) + 3\,H_2(g) \qquad (2)$$

we could find the amount of aluminum in a pure sample by measuring the amount of hydrogen produced by its reaction with an acid solution. In this case two moles of aluminum would produce three moles of hydrogen.

Since the amount of hydrogen produced by a gram of zinc is not the same as the amount produced by a gram of aluminum,

$$1\text{ mol Zn} \rightarrow 1\text{ mol }H_2,\ 65.4\text{ g Zn} \rightarrow 1\text{ mol }H_2,\ 1.00\text{ g Zn} \rightarrow 0.0153\text{ mol }H_2 \qquad (3)$$

$$2\text{ mol Al} \rightarrow 3\text{ mol }H_2,\ 54.0\text{ g Al} \rightarrow 3\text{ mol }H_2,\ 1.00\text{ g Al} \rightarrow 0.0556\text{ mol }H_2 \qquad (4)$$

it is possible to react an alloy of zinc and aluminum of known mass with acid, determine the amount of hydrogen gas evolved, and calculate the percentages of zinc and aluminum in the alloy, using relations (3) and (4). The object of this experiment is to make such an analysis.

In this experiment you will react a weighed sample of an aluminum-zinc alloy with an excess of acid and collect the hydrogen gas evolved over water (Fig. 9.1). If you meas-

*W. L. Masterton, J. Chem. Educ. 38, 558 (1961).

ure the volume, temperature, and total pressure of the gas and use the Ideal Gas Law, taking proper account of the pressure of water vapor in the system, you can calculate the number of moles of hydrogen produced by the sample:

$$P_{H_2} V = n_{H_2} RT, \qquad n_{H_2} = \frac{P_{H_2} V}{RT}; \qquad R = 8.31 \, \frac{kPa \cdot dm^3}{mol \cdot K} \tag{5}$$

The volume V and the temperature T of the hydrogen are easily obtained from the data. The pressure exerted by the dry hydrogen P_{H_2} requires more attention. The total pressure P of gas in the bottle is, by Dalton's Law, equal to the partial pressure of the hydrogen P_{H_2} plus the partial pressure of the water vapor P_{H_2O}:

$$P = P_{H_2} + P_{H_2O} \tag{6}$$

The water vapor in the bottle is present with liquid water, so the gas is saturated with water vapor; the pressure P_{H_2O} under these conditions is equal to the vapor pressure VP_{H_2O} of water at the temperature of the experiment. This value is constant at a given temperature, and will be found in Appendix I at the end of this manual. The total gas pressure P in the flask is very nearly equal to the barometric pressure P_{bar}.[*]

Substituting these values into (6) and solving for P_{H_2}, we obtain

$$P_{H_2} = P_{bar} - VP_{H_2O} \tag{7}$$

Using (5), you can now calculate n_{H_2}, the number of moles of hydrogen produced by your weighed sample. You can then calculate the percentages of Al and Zn in the sample by properly applying (3) and (4) to your results. For a sample containing g_{Al} grams Al and g_{Zn} grams Zn, it follows that

$$n_{H_2} = g_{Al} \times 0.0556 + g_{Zn} \times 0.0153 \tag{8}$$

For a one gram sample, g_{Al} and g_{Zn} represent the mass fractions of Al and Zn, that is, % Al/100 and % Zn/100. Therefore

$$N_{H_2} = \frac{\% \, Al}{100} \times 0.0556 + \frac{\% \, Zn}{100} \times 0.0153 \tag{9}$$

where N_{H_2} = number of moles of H_2 produced *per gram* of sample.

Since it is also true that

$$\% \, Zn = 100 - \% \, Al \tag{10}$$

(9) can be written in the form

$$N_{H_2} = \frac{\% \, Al}{100} \times 0.0556 + \frac{100 - \% \, Al}{100} \times 0.0153 \tag{11}$$

We can solve equation (11) directly for % Al if we know the number of moles of H_2 evolved per gram of sample. To save time in the laboratory and to avoid arithmetic errors, it is highly desirable to prepare in advance a graph giving N_{H_2} as a function of % Al. Then when N_{H_2} has been determined in the experiment, % Al in the sample can be read directly from the graph. Directions for preparing such a graph are given in Problem 1 in the Advance Study Assignment.

---

[*]In principle a small correction should be made for the difference in heights of the water levels inside and outside the sample bottle. In practice the error made by neglecting this effect is much smaller than other experimental errors.

EXPERIMENTAL PROCEDURE

WEAR YOUR SAFETY GLASSES WHILE
PERFORMING THIS EXPERIMENT

Obtain a drying tube and sample of Al-Zn alloy from the stockroom. Assemble the apparatus as shown in Figure 9.1. The top of the funnel should be at least 3 cm higher than the top of the tube leading from the drying tube to the pneumatic trough.

Weigh the vial containing your unknown on the analytical balance. Transfer about half the alloy to a piece of paper and weigh the vial again. The sample transferred should weigh between 0.100 and 0.180 g. When you have a sample of the proper weight, wrap it in some copper wool and place it in the drying tube.

Fill the apparatus by pouring water through the funnel. Close the clamp when all of the air is out of the apparatus. If bubbles of air appear in the tubing after the clamp is closed, check the rubber connections for leaks.

Fill a gas-collecting bottle with water and slide a glass plate across the mouth of the bottle in such a way that no air bubbles are trapped in the bottle. Invert the covered bottle in the trough and remove the glass plate. Insert the glass tube into the bottle.

Pour 10 cm³ of 12 M HCl into the funnel. Open the clamp slowly to allow the acid to come into contact with the metal sample. Close the clamp when the metal starts to give off bubbles vigorously. When the reaction slows down, admit more acid. Be careful not to allow any air bubbles to get into the tube leading from the funnel. The second sample of metal can be weighed while the first sample is being allowed to react completely (approximately 15 min). When gas bubbles are no longer evolved from the metal, fill the funnel with water and open the clamp slowly to flush all the gas out of the tubing and into the collection bottle. *CAUTION:* Be careful when using 12 M HCl; it is corrosive.

Measure the temperature of the water in the trough. Read the atmospheric pressure from the barometer. To measure the volume of gas inside the collection vessel, first cover the mouth of the bottle under the water and invert the bottle. Remove the glass plate and dry the outside of the bottle. Weigh it on a platform balance to ±0.1 g. Fill the bottle completely with water and weigh again. The difference in the masses is the mass of water equivalent to the volume occupied by the gas produced in the reaction. Assuming that the density of water is 1.00 g/cm³, we find that this is numerically equal to the volume of gas in cubic centimetres.

Repeat the experiment using the second weighed sample of metal.

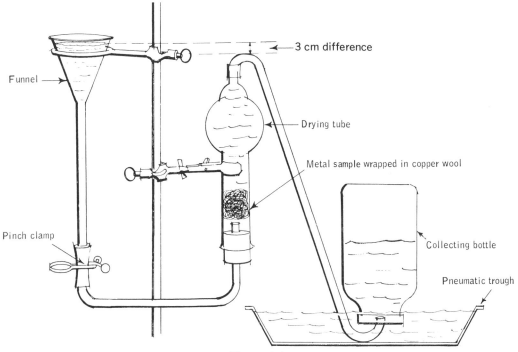

Figure 9.1

DATA: Analysis of an Aluminum-Zinc Alloy

	Trial 1	Trial 2
Mass of sample plus vial	_____ g	
Mass of about one-half the sample plus vial	_____ g	_____ g
Mass of vial plus any remaining sample		_____ g
Temperature of water = temperature of H_2	_____ °C	_____ °C
Barometric pressure (1 mm Hg = 0.1333 kPa)	_____ kPa	_____ kPa
Mass of bottle partially filled with water	_____ g	_____ g
Mass of bottle filled with water	_____ g	_____ g

CALCULATIONS

	Trial 1	Trial 2
Mass of sample (0.10 to 0.18 g)	_____ g	_____ g
Volume of H_2, V	_____ dm³	_____ dm³
Temperature of H_2, T	_____ K	_____ K
Vapor pressure of H_2O at T, VP_{H_2O}, from Appendix I	_____ kPa	_____ kPa
Pressure of dry H_2, P_{H_2} (Eq. 7)	_____ kPa	_____ kPa
Moles H_2 from sample, n_{H_2} (Eq. 5)	_____ mol	_____ mol
Moles H_2 per gram of sample, N_{H_2}	_____ mol/g	_____ mol/g
% Al (read from graph)	_____ %	_____ %

Unknown no.   _____

ADVANCE STUDY ASSIGNMENT: Analysis of an Aluminum-Zinc Alloy

1.   On the following page, construct a graph of N_{H_2} vs. % Al. To do this, refer to Equation (11) and the discussion preceding it. Note that a plot of N_{H_2} vs. % Al should be a straight line (why?). To fix the position of a straight line, it is necessary to locate only two points. The most obvious way to do this is to calculate N_{H_2} when % Al = 0 and when % Al = 100. If you wish, you may also locate intermediate points (for example, N_{H_2} when % Al = 50, and so forth); all these points should be on the same straight line.

2.   A student finds that a sample of an Al-Zn alloy weighing 0.235 g reacts with excess acid to generate 0.185 dm³ of hydrogen, measured over water at a temperature of 22°C and a total pressure of 99.3 kPa. Calculate:

   a. the partial pressure of dry H_2 (P_{H_2})   _____

   b. the number of moles of H_2 evolved (n_{H_2})   _____

   c. the number of moles H_2 per gram of sample _____

   d. % Al, from graph   _____

   e. % Al, from Equation (11)   _____

3.   The nature of the reaction of various metals with excess acid is as follows:

   Mg   24.4 g Mg → 1 mol H_2 Cu   no reaction

   Fe   55.8 g Fe → 1 mol H_2 Hg   no reaction

   Co   59.0 g Co → 1 mol H_2

   On this basis, indicate whether alloys of the following metals could be analyzed satisfactorily by the procedure of this experiment. Indicate your reasoning.

   a. Cu-Fe

   b. Mg-Co

   c. Co-Fe

   d. Cu-Hg

Analysis of an Aluminum-Zinc Alloy
(Advance Study Assignment)

N_{H_2}

0.050

0.040

0.030

0.020

0.010

25%   50%   75%   100%

Per cent Al

EXPERIMENT

10 • Molecular Mass of a Volatile Liquid

One of the important applications of the Ideal Gas Law is found in the experimental determination of the molecular masses of gases and vapors. In order to measure the molecular mass of a gas or vapor we need simply to determine the mass of a given sample of the gas under known conditions of temperature and pressure. If the gas obeys the Ideal Gas Law,

$$PV = nRT \qquad (1)$$

If the pressure P is in kilopascals, the volume V in cubic decimetres, the temperature T in K, and the amount n in moles, then the gas constant R is equal to 8.31 kPa $\cdot$ dm³/mol $\cdot$ K.

The number of moles n is equal to the mass g of the gas divided by its gram molecular mass (GMM). Substituting into (1), we have:

$$PV = \frac{gRT}{GMM} \qquad GMM = \frac{gRT}{PV} \qquad (2)$$

This experiment involves measuring the gram molecular mass of a volatile liquid by using Equation (2). A small amount of the liquid is introduced into a weighed flask. The flask is then placed in boiling water, where the liquid will vaporize completely, driving out the air and filling the flask with vapor at barometric pressure and the temperature of the boiling water. If we cool the flask so that the vapor condenses, we can measure the mass of the vapor and calculate the value for GMM.

EXPERIMENTAL PROCEDURE*

WEAR YOUR SAFETY GLASSES WHILE
PERFORMING THIS EXPERIMENT

Obtain a special round bottom flask, a stopper and cap, and an unknown liquid from the storeroom. Support the flask on an evaporating dish or in a beaker at all times. If you should break or crack the flask, report it to your instructor immediately so that it can be repaired. With the stopper loosely inserted in the neck of the flask, weigh the empty dry flask on the analytical balance. Use a copper loop, if necessary, to suspend the flask from the hook of a two-pan balance.

Pour about half your unknown liquid, about 5 cm³, into the flask. Assemble the apparatus as shown in Figure 10.1. Place the cap on the neck of the flask. Add a few boiling chips to the water in the 600 cm³ beaker and heat the water to the boiling point. Watch the liquid level in your flask; the level should gradually drop as vapor escapes through the cap. After all the liquid has disappeared and no more vapor comes out of the cap, continue to boil the water gently for 5 to 8 min. Measure the temperature of the boiling water. Shut off the burner and wait until the water has stopped boiling (about ½ min) and then loosen the clamp holding the flask in place. Slide out the flask, remove the cap, and *immediately* insert the stopper used previously.

Remove the flask from the beaker of water, holding it by the neck, which will be cooler. Immerse the flask in a beaker of cool water to a depth of about 5 cm. After

---

*See W. L. Masterton and T. R. Williams, J. Chem. Educ. 36, 528, (1959).

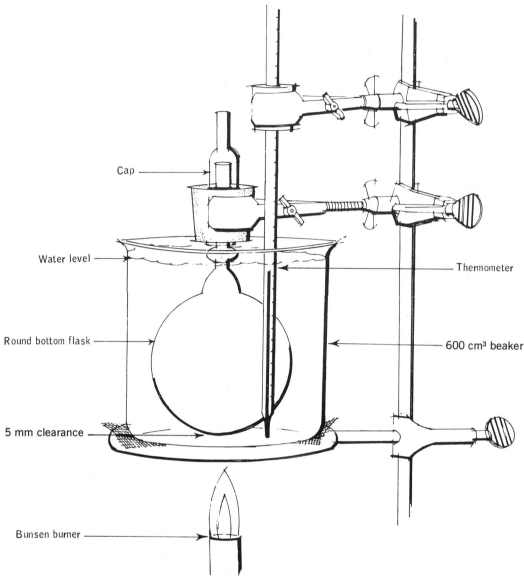

Cap

Water level

Round bottom flask

5 mm clearance

Bunsen burner

Thermometer

600 cm³ beaker

Figure 10.1

holding the flask in the water for about two minutes to allow it to cool, carefully remove the stopper *for not more than a second or two* to allow air to enter, and again insert the stopper. (As the flask cools the vapor inside condenses and the pressure drops, which explains why air rushes in when the stopper is removed.)

Dry the flask with a towel to remove the surface water. Loosen the stopper momentarily to equalize any pressure differences, and reweigh the flask. Read the atmospheric pressure from the barometer.

Repeat the procedure using the other half of your liquid sample.

You may obtain the volume of the flask from your instructor. Alternatively, he may direct you to measure its volume by weighing the flask stoppered and full of water on a rough balance. *Do not* fill the flask with water unless specifically told to do so.

When you have completed the experiment, return the flask to the storeroom; do not attempt to wash or clean it in any way.

DATA: Molecular Mass of a Volatile Liquid

	Trial 1	Trial 2
Unknown no.	_____	
Mass of flask and stopper	_____ g	_____ g
Mass of flask, stopper, and condensed vapor	_____ g	_____ g
Mass of flask, stopper, and water (see directions)	_____ g	_____ g
Temperature of boiling water bath	_____ °C	_____ °C
Barometric pressure (1 mm Hg = 0.1333 kPa)	_____ kPa	_____ kPa

CALCULATIONS AND RESULTS

	Trial 1	Trial 2
Pressure of vapor, P	_____ kPa	_____ kPa
Volume of flask (volume of vapor), V	_____ dm³	_____ dm³
Temperature of vapor, T	_____ K	_____ K
Mass of vapor, g	_____ g	_____ g
Gram molecular mass of unknown, as found by substitution into Equation (2)	_____ g	_____ g

65

ADVANCE STUDY ASSIGNMENT: Molecular Mass of a Volatile Liquid

1.   A 3 cm³ sample of an unknown liquid is vaporized in a flask having a volume of 245 cm³. At 100°C., 0.504 g of the vapor exert a pressure of 99.3 kPa. Calculate the gram molecular mass of the unknown liquid.

_____ g

2.   How would each of the following procedural errors affect the results to be expected in this experiment? Give your reasoning in each case.
   a. All of the liquid was not vaporized before the flask was removed from the water bath.

   b. The water bath never reached the boiling point since the liquid all vaporized before the bath got up to 100°C, and so the student stopped heating.

   c. The barometric pressure of the room decreased between the time the pressure was measured and the time when the vaporization experiment was carried out.

   d. The flask was left in the boiling water bath for an hour instead of for 8 min.

3.   The liquids used in this experiment all have appreciable vapor pressures at room temperature. What effect will this have upon the values of molecular mass obtained?

ATOMIC STRUCTURE: CHEMICAL BONDING

EXPERIMENT

11 • The Atomic Spectrum of Hydrogen

According to the quantum theory, atoms and molecules can exist only in certain states, each of which has an associated fixed amount of energy. When an atom or molecule changes its state, it must absorb or emit an amount of energy equal to the difference between the energy of the initial and final states. This energy may be absorbed or emitted in the form of light, in which case the relationship between the change in energy and the wavelength of the light which is associated with the transition is given by the equation:

$$|\Delta E| = \frac{hc}{\lambda} \tag{1}$$

where $|\Delta E|$ is the absolute value of the change in energy in joules, h is Planck's constant, 6.625×10^{-34} J · s, c is the speed of light, 2.998×10^8 m/s and λ is the wavelength in metres. The change in energy, ΔE, of the atom or molecule is positive if light is absorbed and negative if it is emitted.

Atomic and molecular spectra are the result of changes in energy which occur in atoms and molecules when they are excited by various means. The emission spectrum of an atom gives us information about the spacings between the allowed energy levels of that atom. The different wavelengths present in the light can be used to establish the actual energy levels available to the atom. Conversely, given the set of energy levels for an atom, one can predict its atomic spectrum and determine which levels were involved in any observed line in the spectrum.

Since you are probably not familiar with the way in which Equation 1 is used, let us consider a specific example. On the left side of Figure 11.1, we have shown two of the lowest energy levels in which the sodium atom can exist. These occur at 8.25×10^{-19} J and 4.87×10^{-19} J, respectively, *below* the energy the atom has when it ionizes, which is arbitrarily assigned the value of zero. The actual energies on this basis are both negative, with the lower energy having the more negative value.

Ordinarily a sodium atom will exist in its lowest possible energy state, which is called its ground state. If the atom is excited, say in a flame, to its next higher state, it will be unstable and will very quickly make a transition back down to its ground state, as indicated by the arrow in Figure 11.1. In making the transition, the energy of the atom will decrease by about 3.38×10^{-19} J. This amount of energy may be radiated as light, which will have a wavelength given by Equation 1. Below the left side of the figure, we have calculated that wavelength which turns out to be 5.88×10^{-7} m. Wavelengths of light are ordinarily given in nanometres; since 1 nm = 10^{-9} m, the wavelength is 588 nm. Atomic spectra arise from transitions of this sort, and the wavelengths associated with those transitions can all be calculated by the method we have used here.

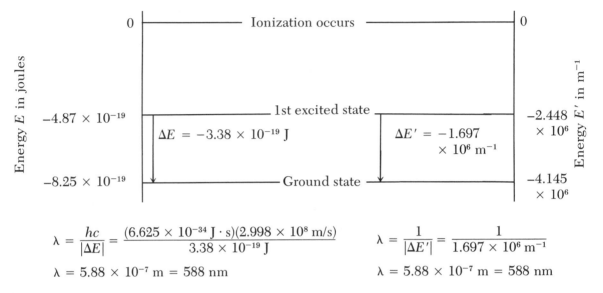

$$\lambda = \frac{hc}{|\Delta E|} = \frac{(6.625 \times 10^{-34} \text{ J} \cdot \text{s})(2.998 \times 10^{8} \text{ m/s})}{3.38 \times 10^{-19} \text{ J}} \qquad \lambda = \frac{1}{|\Delta E'|} = \frac{1}{1.697 \times 10^{6} \text{ m}^{-1}}$$

$$\lambda = 5.88 \times 10^{-7} \text{ m} = 588 \text{ nm} \qquad\qquad \lambda = 5.88 \times 10^{-7} \text{ m} = 588 \text{ nm}$$

Figure 11.1.   Calculations of a Wavelength in the Atomic Spectrum of Sodium

Many years ago spectroscopists recognized that wavelength calculations would be simplified if the energy in Equation 1 were expressed not in joules, but in units of joules/hc, which turn out to have the dimensions of m^{-1} and are called wavenumbers, or reciprocal metres. Such energies we will call E', and on the right side of Figure 11.1, we have shown the same energy levels of the Na atom on that basis, with energies E' in m^{-1}. The advantage of expressing E' in m^{-1} is that Equation 1, written in terms of E', becomes simply

$$\frac{|\Delta E|}{hc} = |\Delta E'| = \frac{1}{\lambda} \qquad \text{and} \qquad \lambda = \frac{1}{|\Delta E'|} \qquad (2)$$

To find the wavelength in metres for a transition, one need only take the reciprocal of $\Delta E'$. The calculation of λ by this approach is shown on the right side of Figure 11.1. You can see that it gives the same result as that obtained previously, but the mathematics required is somewhat simpler.

The simplest atomic spectrum is that of the hydrogen atom. In 1886 Balmer showed that the lines in the spectrum of the hydrogen atom had wavelengths that could be expressed by a rather simple equation. Bohr, in 1913, explained the spectrum on a theoretical basis with his famous model of the hydrogen atom. According to Bohr's theory, the energies E_n' allowed to an H atom are all given by the following equation:

$$E_n' = -\frac{R}{n^2} \qquad (3)$$

where R is a constant predicted by the theory and n is an integer, 1, 2, 3,. . ., called a quantum number. It has been found that all the lines in the atomic spectrum of hydrogen can be associated with energy levels in the atom which are predicted with great accuracy by Bohr's equation.

In this experiment we will furnish you with the wavelengths of some of the observed lines in the hydrogen atomic spectrum and ask you to explain the origin of each line in terms of the energy levels of the atom. The wavelengths to be considered are given in Table 11.1.

TABLE 11.1   SOME WAVELENGTHS (IN NANOMETRES) IN THE SPECTRUM OF THE HYDROGEN ATOM AS MEASURED IN A VACUUM

Wavelength	Assignment $n_{hi} \longrightarrow n_{lo}$	Wavelength	Assignment $n_{hi} \longrightarrow n_{lo}$	Wavelength	Assignment $n_{hi} \longrightarrow n_{lo}$
97.25	_____	410.29	_____	1005.3	_____
102.57	_____	434.17	_____	1094.1	_____
121.57	_____	486.27	_____	1282.2	_____
389.02	_____	656.47	_____	1875.6	_____
397.14	_____	954.86	_____	4050	_____

EXPERIMENTAL PROCEDURE

There are several ways in which one might analyze an atomic spectrum, given the energy levels of the atom, but a simple and powerful one is to calculate the wavelengths of some of the lines that are allowed and to see if they match those which are observed. We shall use this method in our experiment.

The value of R in Equation 3 can be measured very accurately and is found to be 1.0967758×10^7 m^{-1}, perhaps the most accurately known of all physical constants. Since we will be working with electronic calculators, we can use most, or all, of the available precision. On this basis, Equation 3 takes the form

$$E'_n = -\frac{1.0967758 \times 10^7}{n^2} \text{ m}^{-1} \tag{4}$$

A. Calculations of the Energy Levels of the Hydrogen Atom.   Given the expression for E'_n in Equation 4, it is possible to calculate the energy for each of the allowed levels of the H atom starting with $n = 1$. Using your calculator, calculate the energy in m^{-1} of each of the ten lowest levels of the H atom. Note that the energies are all negative, so that the *lowest* energy will have the *largest* allowed negative value. Enter these values in the table of energy levels, Table 11.2. On the graph paper provided, plot along the y axis each of the six lowest energies, drawing a horizontal line at the allowed level and writing the value of the energy alongside the line near the y axis. Write the quantum number associated with the level to the right of the line.

B. Calculation of the Wavelength of the Lines in the Hydrogen Spectrum.   The lines in the hydrogen spectrum all arise from jumps made by the atom from one energy level to another. The wavelengths in metres of these lines can be calculated by Equation 2, where $\Delta E'$ is the difference in energy between any two allowed levels. For example, to find the wavelength of the spectral line associated with a transition from the $n = 2$ level to the $n = 1$ level, calculate the difference, $\Delta E'$, between the energies of those two levels.

The reciprocal of $\Delta E'$ will, by Equation 2, be the wavelength in metres of the spectral line. To convert that to nanometres, use the conversion factor, 1 nm = 10^{-9} m.

Using the procedure we have outlined, calculate the wavelengths (nm) of all the lines we have indicated in Table 11.3. That is, calculate the wavelengths of all the lines that can arise from transitions between any two of the six lowest levels of the H atom. Enter these values in Table 11.3.

C. Assignment of Observed Lines in the Hydrogen Spectrum.   Compare the wave lengths you have calculated with those which are listed in Table 11.1. If you have made your calculations properly, your wavelengths should match, within the error of your calculation, several of those which are observed. On the line opposite each wavelength in Table 11.1, write the quantum numbers of the upper and lower states for each line whose origin you can recognize by comparison of your calculated values with the observed values. On the energy level diagram, draw a vertical arrow pointing down (light is emitted, $\Delta E' < 0$) between those pairs of levels which you associate with any of the observed wavelengths. By each arrow write the wavelength of the line originating from that transition.

There are a few wavelengths in Table 11.1 which have not yet been calculated. By assignments already made and by an examination of the transitions you have marked on the graph, deduce the quantum states that are likely to be associated with one of the as yet unassigned lines. Calculate the wavelength for the transition between those states. When you have matched a calculated with an observed wavelength, write the associated quantum numbers as before in Table 11.1; continue until all the lines in the table have been assigned.

D. The Balmer Series.   This is the most famous series in the atomic spectrum of hydrogen. Carry out calculations in connection with this series as directed in the Data and Calculations section.

DATA AND CALCULATIONS: The Atomic Spectrum of Hydrogen

A.   The Energy Levels of the Hydrogen Atom

Energies are to be calculated from Equation 4 for the ten lowest energy states.

TABLE 11.2

Quantum Number	Energy, E_n', in m^{-1}	Quantum Number	Energy, E_n', in m^{-1}
_____	_____	_____	_____
_____	_____	_____	_____
_____	_____	_____	_____
_____	_____	_____	_____
_____	_____	_____	_____

B.   Calculation of Wavelengths in the Spectrum of the H Atom

TABLE 11.3

n_{higher}	6	5	4	3	2	1

n_{lower}

$$\Delta E' = E_{n_{\text{hi}}}' - E_{n_{\text{lo}}}' = \frac{1}{\lambda}\ m^{-1}$$

$$\lambda(m) = \frac{1}{\Delta E'}$$

$$\lambda(nm) = \lambda(m) \times 1 \times 10^{9}$$

In the upper half of each box write $\Delta E'$, the difference in energy in m^{-1} between $E_{n_{\text{hi}}}'$ and $E_{n_{\text{lo}}}'$. In the lower half of the box, write λ (nm) associated with that value of $\Delta E'$.

Continued on following page

C. Assignment of Wavelengths

A. As directed in the procedure, assign n_{hi} and n_{lo} for each wavelength in Table 11.1 which corresponds to a wavelength calculated in Table 11.3.

B. List below any wavelengths you cannot yet assign and find their origin.

Wavelength, λ observed	Probable transition $n_{hi} \longrightarrow n_{lo}$	ΔE′ transition	λ calculated (nm)
_____	_____	_____	_____
_____	_____	_____	_____
_____	_____	_____	_____
_____	_____	_____	_____

D. 1. THE BALMER SERIES. When Balmer found his famous series for hydrogen in 1886, he was limited experimentally to wavelengths in the visible and near ultraviolet regions from 250 nm to 700 nm. From the entries in Table 11.3, what common characteristic do the lines in the Balmer Series have?

What would be the longest possible wavelength for a line in the Balmer series?

λ = _____ nm

What would be the shortest possible wavelength that a line in the Balmer series could have?

λ = _____ nm

Fundamentally, why must all the lines in the hydrogen spectrum between 250 nm and 700 nm belong to the Balmer series?

Continued on following page

2. THE IONIZATION ENERGY OF HYDROGEN. In a normal hydrogen atom, the electron is in the lowest energy state. The maximum energy that a hydrogen atom can have is 0 m^{-1}, at which point the electron is essentially removed from the atom, and ionization occurs. How much energy in m^{-1} does it take to ionize an H atom?

_____m^{-1}

a. Using the equation: $\Delta E = hc(\Delta E')$, where $\Delta E'$ is the value just calculated, obtain ΔE, the ionization energy in joules of the H atom ($h = 6.625 \times 10^{-34}$ J $\cdot$ s, $c = 2.998 \times 10^8$ m/s).

_____J

b. Calculate the ionization energy, in joules, of a mole of hydrogen atoms (1 mol = 6.022×10^{23} H atoms).

_____J

The Spectrum of Hydrogen

(Data and Calculations)

Energy in m^{-1}

0
-1 × 10^6
-2 × 10^6
-3 × 10^6
-4 × 10^6
-5 × 10^6
-6 × 10^6
-7 × 10^6
-8 × 10^6
-9 × 10^6
-10 × 10^6
-11 × 10^6

ADVANCE STUDY ASSIGNMENT: The Atomic Spectrum of Hydrogen

1. The lowest energy level of the lithium atom lies 4.3487×10^6 m^{-1} below the energy required to ionize the atom. The first excited state of the Li atom is at 2.8579×10^6 m^{-1} below the ionization energy. What wavelength of light is emitted when an Li atom makes a transition from the first excited state to the ground state? What is the ionization energy of the Li atom in m^{-1}? in joules? What is the ionization energy of a mole of Li atoms?

_____ nm

_____ m^{-1}

_____ J

_____ J/mol

2. In the HCl molecule, the energy levels associated with rotation are given by the equation, $E' = 1040\,J\,(J + 1)$ m^{-1}, where J is a quantum number which may have the values 0, 1, 2, ... Find the wavelength of the light which would be absorbed by the molecule in making a transition from the $J = 5$ to the $J = 6$ state.

_____ nm

12 • The Alkaline Earths and the Halogens—Two Families in the Periodic Table

The periodic table arranges the elements in order of increasing atomic number in horizontal rows of such length that elements with similar properties recur periodically, i.e., they fall directly beneath each other in the table. The elements in a given vertical column are referred to as a family. By noting the gradual trends in properties of the members of a family, it is possible to arrange them in the order in which they fall in the periodic table. This is what you will be asked to do in this experiment for two particular families.

The families to be studied are the alkaline earths, Group II A, and the halogens, Group VII A. The alkaline earths are all active metals and include barium, beryllium, calcium, magnesium, radium, and strontium. Beryllium compounds, rarely encountered, are often very poisonous, and radium is highly radioactive, so we shall not include these two elements in the experiment. The elements in the halogen family are astatine, bromine, chlorine, fluorine, and iodine. Of these we will omit astatine, which is radioactive, and fluorine, which is the most chemically reactive of all the elements and somewhat dangerous to work with.

The experiments with the alkaline earths involve determining the relative solubilities of the salts formed by the alkaline earth cations with sulfate, carbonate, oxalate, and chromate ions. When solutions containing these M^{2+} cations are mixed with the above X^{2-} anions, the following reaction will occur if the salt MX is not very soluble:

$$M^{2+}(aq) + X^{2-}(aq) \longrightarrow MX(s) \tag{1}$$

$$M^{2+} = Ba^{2+}, Ca^{2+}, Mg^{2+}, \text{ or } Sr^{2+} ; X^{2-} = SO_4^{2-}, CO_3^{2-}, C_2O_4^{2-}, \text{ or } CrO_4^{2-}$$

The trends in solubilities of these salts are consistent with the order of the II A elements in the periodic table and can be used to establish that order.

The elementary halogens are all oxidizing agents, which means that they tend to react with other substances in such a way as to gain electrons; the reaction is called an oxidation-reduction reaction and results in the halogen (X_2) being reduced to a halide anion (X^-). Since the oxidizing powers of the elementary halogens are not the same, if we mix a solution of the halogen X_2 with a solution containing a halide ion Y^-, the following reaction may occur:

$$X_2(aq) + 2Y^-(aq) \longrightarrow Y_2(aq) + 2X^-(aq) \tag{2}$$

The reaction will occur if X_2 is a better oxidizing agent than Y_2, since then X_2 can produce Y_2 by removing electrons from the Y^- ion. Conversely, if Y_2 is a stronger oxidizing agent than X_2, reaction 2 will not proceed as written, but will be spontaneous in the opposite direction.

We will test the oxidizing powers of the halogens using this approach. You will be able to tell whether a reaction proceeds by observing the colors of the solutions. The halogens have characteristic colors in water and, particularly, in 1,1,1-trichloroethane,

CCl_3CH_3 (TCE), and these will facilitate your tests. For example, Br_2 in solution in TCE has a reddish brown color, quite different from that of either Cl_2 or I_2 in TCE solution. (Bromide ion, Br^-, like all the halide ions, is colorless.) If we shake a solution of bromine water with TCE, most of the Br_2 will go into the TCE phase and impart its color to it. Then, if we add a solution of a salt containing an excess of another halide ion, say Cl^-, the following oxidation-reduction reaction may or may not occur:

$$Br_2(aq) + 2\,Cl^-(aq) \rightarrow Cl_2(aq) + 2\,Br^-(aq) \tag{3}$$

If the reaction occurs, the Br_2 color in the TCE will be replaced by that of Cl_2, whereas if it does not occur, there will be no appreciable color change. If, indeed, you observe a color change, indicating a reaction occurs, then you can say that Br_2 is a stronger oxidizing agent than Cl_2, since it can produce Cl_2 by oxidizing Cl^- ion. In the event no reaction occurs, as implied by the fact that the color of the TCE remains substantially unchanged, then Cl_2 is a stronger oxidizing agent than Br_2. Using this approach with the various possible mixtures of halogens and halide ions, it is quite easy to arrange the halogens in order of increasing oxidizing power.

We will also investigate the relative solubilities of the salts formed between the silver ion and the various halide ions. If a solution of silver nitrate, $AgNO_3$, is added to a solution of a halide salt, MX, the following reaction will occur:

$$Ag^+(aq) + X^-(aq) \longrightarrow AgX(s) \tag{4}$$

An insoluble precipitate of AgX forms immediately. This precipitate, although it is very insoluble in water, may dissolve in NH_3 solutions, since there the following reaction tends to occur:

$$AgX(s) + 2\,NH_3(aq) \longrightarrow Ag(NH_3)_2{}^+(aq) + X^-(aq) \tag{5}$$

The tendency for reaction 5 to proceed increases with increasing NH_3 concentration and with the increasing solubility of the silver halide.

The oxidizing powers of the halogens and the solubilities of their silver salts in NH_3 solutions will allow you to arrange the halogens in the order in which they should appear in the periodic table.

Given the properties of the alkaline earths and the halogens as observed in this experiment, it is possible to develop a systematic procedure for determining the presence of any II A cation and any given halide ion in a solution. In the last part of the experiment you will be asked to set up such a procedure and use it to establish the identity of an unknown solution containing a single alkaline earth halide.

EXPERIMENTAL PROCEDURE

WEAR YOUR SAFETY GLASSES WHILE
PERFORMING THIS EXPERIMENT

I. Relative Solubilities of Some Salts of the Alkaline Earths.   Add about 1 cm³ (approximately 12 drops) of 0.1 M solutions of the nitrate salts of barium, calcium, magnesium, and strontium to separate small test tubes. To each tube add 1 cm³ of 1 M H_2SO_4 and stir with your glass stirring rod. (Rinse your stirring rod in a beaker of water between tests.) Record your results in the table, noting whether a precipitate forms, as well as any characteristics that might distinguish it.

Repeat the experiment using 1 M Na_2CO_3 as the precipitating reagent and record your observations. Then test for the solubilities of the oxalate salts with 0.25 M $(NH_4)_2C_2O_4$. Finally, determine the relative solubilities of the chromate salts, using 1 cm³ of 1 M K_2CrO_4 plus 1 cm³ of 1 M acetic acid as the testing reagent.

II. Relative Oxidizing Powers of the Halogens.   In a small test tube place 2 cm³ of bromine-saturated water and add 1 cm³ of trichloroethane. Stopper the test tube and shake until the bromine color is mostly in the TCE layer. *Caution:* Don't use your finger to stopper the tube, since halogen solutions can give you a bad chemical burn. Repeat the experiment using chlorine water and iodine water, noting any color changes as the bromine, chlorine, and iodine are extracted from the water into the TCE.

Shake 1 cm³ of bromine water with 1 cm³ of TCE in a small test tube. Add 1 cm³ 0.1 M NaCl solution, stopper, and shake. Using another sample of Br_2 solution with TCE, repeat the experiment using 0.1 M NaI solution. In each case observe the color of the TCE phase before and after addition of the halide to determine whether an oxidation-reduction reaction has occurred. Repeat the tests on 1 cm³ samples of the three halide solutions using chlorine water and then iodine water in TCE. Since mixtures of Cl^- with Cl_2 and I^- with I_2 will not aid in deciding on relative oxidizing powers, you will not need to test them. Record your observations for each test.

III. Solubilities of Silver Halide Salts. Add 1 cm³ of 0.1 M solutions of the three sodium halides to separate test tubes. Add a few drops of 0.1 M $AgNO_3$ to each test tube and stir. Note the color of each precipitate. Let the precipitates settle, centrifuge if necessary, and pour off the liquid. To the precipitate add 6 M NH_3 drop-wise with stirring, noting the solubility in each case. With any precipitates which do not dissolve after addition of about 2 cm³ of ammonia, pour off the 6 M NH_3 and test for solubility of the solid in 15 M NH_3. Record your results.

IV. Identification of Unknown Salt. On the basis of your observations, devise a scheme by which you can establish which alkaline earth cation and which halide ion is present in a solution containing a single alkaline earth halide. Use your procedure to identify the salt in your unknown solution.

DATA AND OBSERVATIONS: Periodicity of Chemical Properties

I. Solubilities of Salts of the Alkaline Earths

	1 M H_2SO_4	1 M Na_2CO_3	0.25 M $(NH_4)_2C_2O_4$	1 M K_2CrO_4 1 M Acetic Acid
$Ba(NO_3)_2$				
$Ca(NO_3)_2$				
$Mg(NO_3)_2$				
$Sr(NO_3)_2$				

P = precipitate forms; S = no precipitate.

Note any distinguishing characteristics of ppt.

Consider the relative solubilities of the Group II A cations in the various precipitating reagents. On the basis of the trends you observed, list the four alkaline earths in the order in which they should appear in the periodic table. *Start with the one which forms the most soluble oxalate.*

_____ _____ _____ _____

Why did you arrange the elements as you did? Is the order consistent with the properties of the cations in all of the precipitating reagents?

II. Relative Oxidizing Powers of the Halogens
   A.   Color of the halogen in solution:

	Br_2	Cl_2	I_2
Water	_____	_____	_____
TCE	_____	_____	_____

Continued on following page   **83**

B. Reactions between halogens and halides:

	Br⁻	Cl⁻	I⁻
Br₂			
Cl₂			
I₂			

State initial and final colors of TCE layer. R = reaction occurs; NR = no reaction occurs.

III. Properties of Silver Halide Salts

	AgBr	AgCl	AgI
1. Color	_____	_____	_____
2. Solubility in NH_3 solution	_____	_____	_____

S = soluble in 6 M NH_3; SS = soluble in 15 M NH_3; IS = insoluble in NH_3.

On the basis of trends in oxidizing power and in solubility of the silver halide salts in NH_3 solutions, arrange the halogens in the order in which they should be listed in the periodic table. Start with the strongest oxidizing agent.

_____   _____   _____

IV. Given a solution known to contain one Group II A cation and one VII A anion, devise a scheme, based on the properties of these ions as you observed them in this experiment, which would allow you to determine which cation and which anion are present.

Continued on following page

Use your scheme to analyze your unknown solution.
Observations:

Cation present _____ Anion present _____

Unknown No. _____

ADVANCE STUDY ASSIGNMENT: Periodic Properties of Substances

1.   Calcium hydroxide is slightly soluble (2 g/dm^3) in water, whereas barium hydroxide is moderately soluble (28 g/dm^3). Would you expect magnesium hydroxide to be more soluble or less soluble than strontium hydroxide? Why?

2.   Substances A, B, and C can all behave as oxidizing agents. When a solution of substance A is mixed with a solution containing B$^-$ ions, no reaction occurs. When a solution of A is mixed with a solution containing C$^-$ ions, substance C is produced. Arrange A, B, and C in order of increasing strength as oxidizing agents.

3.   Nitrite ion, NO$_2^-$, is a good oxidizing agent in acid solution, as is elementary iodine, I$_2$. In aqueous solution, NO$_2^-$ is colorless, while I$_2$ is brown. When NO$_2^-$ is added to an acidic solution of KI, the solution turns brown. Which is the stronger oxidizing agent under these conditions, NO$_2^-$ or I$_2$? State your reasoning.

13 • The Geometrical Structure of Molecules: An Experiment Using Molecular Models

Many years ago it was observed that in many of its compounds the carbon atom formed four chemical linkages to other atoms. As early as 1870, graphic formulas of carbon compounds were drawn as shown:

$$
\begin{array}{ccc}
& H & \\
& | & \\
H\!-\!\!&C&\!\!-\!H \\
& | & \\
& H &
\end{array}
\qquad\qquad
\begin{array}{ccc}
H & & H \\
| & & | \\
C & \!\!=\!\! & C \\
| & & | \\
H & & H
\end{array}
$$

methane   ethylene

Although such drawings as these would imply that the atom-atom linkages, indicated by valence strokes, lie in a plane, chemical evidence, particularly the existence of only one substance with the graphic formula

$$
\begin{array}{ccc}
& Cl & \\
& | & \\
H\!-\!\!&C&\!\!-\!Cl \\
& | & \\
& H &
\end{array}
$$

requires that the linkages be directed toward the corners of a tetrahedron, at the center of which is the carbon atom.

The concept of a tetrahedral carbon atom was developed and used extensively by organic chemists during the latter part of the nineteenth century. If carbon atoms are considered to be represented by tetrahedra in the manner indicated, single carbon-carbon bonds arise when two such tetrahedra share a common corner, double bonds arise when the tetrahedra share an edge, and triple bonds arise when the tetrahedra share a face. Long before physical methods for confirmation were available, the model correctly predicted that ethylene, $H_2C=CH_2$, would be a planar molecule and that acetylene, $HC\equiv CH$, would be linear.

The physical significance of the chemical linkages between atoms, expressed by the lines or valence strokes in molecular structure diagrams, became evident soon after the discovery of the electron. In 1916 in a classic paper, G. N. Lewis suggested, on the basis of chemical evidence, that the single bonds in graphic formulas involve two electrons and that an atom tends to hold eight electrons in its outermost or valence shell.

Lewis' proposal that atoms generally have eight electrons in their outer shell proved to be extremely useful and has come to be known as the octet rule. It can be applied to many atoms, but is particularly important in the treatment of covalent compounds of atoms in the second row of the periodic table. For atoms such as carbon, oxygen, nitrogen, and fluorine, the eight valence electrons occur in pairs that occupy tetrahedral positions around the central atom core. Some of the electron pairs do not participate directly in chemical bonding and are called unshared or nonbonding pairs; however, the structures of compounds containing such unshared pairs reflect the tetrahedral arrangement of the four pairs of valence shell electrons. In the H_2O molecule, which obeys the octet rule, the four pairs of electrons around the central oxygen atom occupy essentially tetrahedral positions; there are two unshared nonbonding pairs and two bonding pairs which are shared by the O atom and the two H atoms. The H — O — H bond angle is nearly but not exactly tetrahedral since the properties of shared and unshared pairs of electrons are not exactly alike.

For some molecules with a given molecular formula, it is possible to satisfy the octet rule with different atomic arrangements. A simple example would be

The two molecules are called isomers of each other, and the phenomenon is called isomerism. Although the molecular formulas of both substances are the same, C_2H_6O, their properties differ markedly because of their different atomic arrangements.

Isomerism is very common, particularly in organic chemistry, and when double bonds are present, isomerism can occur in very small molecules:

The first two isomers result from the fact that there is no rotation around a double bond, although such rotation can occur around single bonds. The third isomeric structure cannot be converted to either of the first two without breaking bonds.

With certain molecules, given the atomic geometry, it is possible to satisfy the octet rule with more than one bonding arrangement. The classic example is benzene, whose molecular formula is C_6H_6:

These two structures are called resonance structures, and molecules such as benzene, which have two or more resonance structures, are said to exhibit resonance. The actual bonding in such molecules is thought to be an average of the bonding present in the resonance structures. The stability of molecules exhibiting resonance is found to be higher than that anticipated for any single resonance structure.

Once the symmetry of a species has been determined, it is possible to predict its polarity, that is, whether the molecule will contain a region of positive charge and a region of negative charge, and so have a dipole moment. Covalent bonds between different kinds of atoms in molecules are typically polar; all heteronuclear diatomic molecules are polar. In some molecules the polarity from one bond may be cancelled by that arising from others, so that the overall molecular polarity may vanish. Carbon dioxide CO_2, which is linear, is a nonpolar molecule; methane CH_4, which is tetrahedral, is also nonpolar. On the other hand, the related molecules of lower symmetry, COS and CH_3Cl, do not have complete cancellation of bond polarities and are therefore polar.

In this experiment, assuming that all atoms present in the species studied obey the octet rule, you will assemble models of some simple molecules and ions. On the basis of the models you will be able to draw electron dot diagrams and predict the geometrical structure of each species, the existence of isomers, the polarity of the species, and whether resonance structures would be likely to occur.

EXPERIMENTAL PROCEDURE

In this experiment you may work in pairs during the first portion of the laboratory period.

The models you will use consist of drilled wooden balls, short sticks, and springs. The balls represent atomic nuclei surrounded by the inner electron shells. The sticks and springs represent electron pairs and fit in the holes in the wooden balls. The model (molecule or ion) consists of wooden balls (atoms) connected by sticks or springs (chemical bonds). Some sticks may be connected to only one atom (non-bonding pairs).

In this experiment we will deal with atoms that obey the octet rule; such atoms have four electron pairs around the central core and will be represented by black or blue balls with four tetrahedral holes in which there are four sticks or springs. The only exception will be hydrogen atoms, which share two electrons in covalent compounds, and which will be represented by yellow balls with a single hole in which there is a single stick.

In assembling a molecular model of the kind we are considering, it is possible, indeed desirable, to proceed in a systematic manner. We will illustrate the recommended procedure by developing a model for a molecule with the formula CH_2O.

1. Draw electron dot diagrams for each atom in the molecule, letting dots represent valence electrons and the element symbols represent the atomic cores.

For carbon atoms: electron configuration $1s^2 2s^2 2p^2$

$$\text{four valence electrons} \quad \cdot \overset{\displaystyle \cdot}{\underset{\displaystyle \cdot}{C}} \cdot$$

For hydrogen atoms: electron configuration $1s$

$$\text{one valence electron} \quad H \cdot$$

For oxygen atoms: electron configuration $1s^2 2s^2 2p^4$

$$\text{six valence electrons} \quad \cdot \overset{\displaystyle \cdot \cdot}{\underset{\displaystyle \cdot}{O}} \colon$$

Add up the valence electrons for all the atoms in the molecule. In this case there are 12 (four from the C atom, two from the two H atoms, and six from the O atom). If the particle is an ion, add one electron for each negative charge or subtract one for each positive charge on the ion.

2. Select wooden balls and sticks to represent the atoms and electron pairs in the molecule. You might use a black ball for the carbon atom core, a blue ball for the oxygen atom core, and yellow balls to represent the hydrogen atoms. Since there are 12 valence electrons in the molecule and electrons occur in pairs, you will need six sticks to represent the six electron pairs. The sticks will serve both as bonds between atoms and as nonbonding electron pairs.

3. Connect the balls with some of the sticks. (Assemble a skeleton structure for the molecule, joining atoms by single bonds.) In some cases this can only be done in one way. Usually, however, there are various possibilities, some of which are more reasonable than others. In CH_2O the model can be assembled by connecting the two yellow balls (H atoms) to the black ball (C atom) with two of the available sticks, and then using a third stick to connect the black ball to the blue one (O atom).

4. The next step is to use the sticks that are left over in such a way as to fill all the remaining holes in the balls. (Distribute the electron pairs so as to give each atom eight electrons and so satisfy the octet rule.) In the model we have assembled, there is one unfilled hole in the black ball, three unfilled holes in the blue ball, and three available sticks. An obvious way to meet the required condition is to use two sticks to fill two of the holes in the blue ball, and then use two springs instead of two sticks to connect the blue and black balls. The model as completed is shown in Figure 13.1.

5. Interpret the model in terms of the atoms and bonds represented. The sticks and spatial arrangement of the balls will closely correspond to the electronic and atomic arrangement in the molecule. Given our model, we would describe the CH_2O molecule as being planar with single bonds between carbon and hydrogen atoms and a double bond between the C and O atoms. The H—C—H angle is approximately tetrahedral. There are two nonbonding electron pairs on the O atom. Since all bonds are polar and the molecular symmetry does not cancel the polarity in CH_2O, the molecule is polar. The bonding sketch showing electronic structure is given below:

The drawing is really an electron dot structure, with each bond representing two electrons.

(The compound having molecules with the formula CH_2O is well-known and is called formaldehyde. The bonding and structure in CH_2O are as given by the model.)

6. Investigate the possibility of the existence of isomers or resonance structures in the model. It turns out that in the case of CH_2O one can easily construct an isomeric form which obeys the octet rule, in which the central atom is oxygen rather than carbon. It is found that this isomeric form of CH_2O does not exist in nature. As a general rule, however, carbon atoms are almost never found at the end of a chain of atoms; put another way, nonbonding electron pairs on carbon atoms are very rare. Another useful rule of a similar nature is that if there are several oxygen atoms in a simple species containing one other atom, each oxygen atom is attached to that other atom. In the SO_4^{2-} ion, for example, the oxygen atoms are all chemically bound to the sulfur atom. Only rarely do oxygen atoms bond to one another, forming compounds known as peroxides.

Resonance structures are reasonably common. For resonance to occur, however, the atomic arrangement must remain fixed for two or more possible electronic structures. For CH_2O there are no resonance structures.

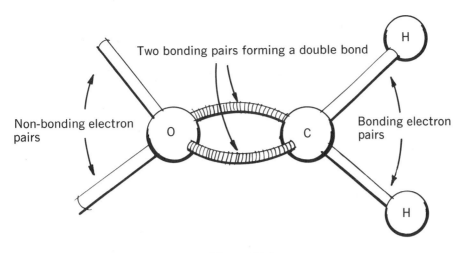

Figure 13.1

A. Using the procedure we have outlined, construct and report on models of the molecules and ions listed here and/or other species assigned by your instructor.

CH_4	H_3O^+	N_2	C_2H_2
CH_2Cl_2	HF	P_4	SO_2
CH_4O	NH_3	C_2H_4	SO_4^{2-}
H_2O	H_2O_2	$C_2H_2Br_2$	CO_2

B. Assuming that stability requires that each atom obey the octet rule, predict the stability of the following species:

$$PCl_3 \qquad CH_3 \qquad OH^- \qquad CO$$

C. When you have completed parts A and B, see your laboratory instructor, who will check your results and assign you a set of unknown species. Working now by yourself, assemble models for each species as in the previous section, and report on the geometry and bonding in each of the unknown species on the basis of the model you construct. Also consider and report on the polarity and the likelihood of existence of isomers and resonance structures for each species.

REPORT: **Geometrical Structures of Molecules Using Molecular Models**

A. Species

Species	Bonding Sketch	Geometry	Isomers or Resonance	Polarity
CH_4				
CH_2Cl_2				
CH_4O				
H_2O				
H_3O^+				

Species	Bonding Sketch	Geometry	Isomers or Resonance	Polarity
HF				
NH_3				
H_2O_2				
N_2				
P_4				

Continued on following page

95

Continued

Species	Bonding Sketch	Geometry	Isomers or Resonance	Polarity
C_2H_4				
$C_2H_2Br_2$				
C_2H_2				

Species	Bonding Sketch	Geometry	Isomers or Resonance	Polarity
SO_2				
SO_4^{2-}				
CO_2				

B. Stability predicted for PCl_3 _____ CH_3 _____ OH^- _____ CO _____

C. Unknowns

_____   _____   _____

_____   _____   _____

EXPERIMENT
14 • Classification of Chemical Substances

Depending on the kind of bonding present in a chemical substance, the substance may be called ionic, molecular, or metallic.

In a solid ionic compound there are ions; the large electrostatic forces between the positively and negatively charged ions are responsible for the bonding which holds these particles together.

In a molecular substance the bonding is caused by the sharing of electrons by atoms. When the stable aggregates resulting from covalent bonding contain relatively small numbers of atoms, they are called molecules. If the aggregates are very large and include essentially all the atoms in a macroscopic particle, the substance is called macromolecular.

Metals are characterized by a kind of bonding in which the electrons are much freer to move than in other kinds of substances. The metallic bond is stable but is probably less localized than other bonds.

The terms ionic, molecular, macromolecular, and metallic are somewhat arbitrary, and some substances have properties that would place them in a borderline category, somewhere intermediate between one group and another. It is useful, however, to consider some of the general characteristics of typical ionic, molecular, macromolecular, and metallic substances, since many very common substances can be readily assigned to one category or another.

IONIC SUBSTANCES

Ionic substances are all solids at room temperature. They are typically crystalline, but may exist as fine powders as well as clearly defined crystals. While many ionic substances are stable up to their melting points, some decompose on heating. It is very common for an ionic crystal to release loosely bound water of hydration at temperatures below 200°C. Anhydrous (dehydrated) ionic compounds have high melting points, usually above 300°C but below 1000°C. They are not readily volatilized and boil at only very high temperatures (Table 14.1).

When molten, ionic compounds conduct an electric current. In the solid state they do not conduct electricity. The conductivity in the molten liquid is attributed to the freedom of motion of the ions, which arises when the crystal lattice is no longer present.

Ionic substances are frequently but not always appreciably soluble in water. The solutions produced conduct the electric current rather well. The conductivity of a solution of a slightly soluble ionic substance is often several times that of the solvent water. Ionic substances are usually not nearly so soluble in other liquids as they are in water. For a liquid to be a good solvent for ionic compounds it must be highly polar, containing molecules with well-defined positive and negative regions with which the ions can interact.

TABLE 14.1   PHYSICAL PROPERTIES OF SOME REPRESENTATIVE
CHEMICAL SUBSTANCES

Substance	M.P.,°C	B.P.,°C	Solubility Water	Solubility Toluene	Electrical Conductance	Classification
NaCl	801	1413	Sol	Insol	High in melt and in soln	Ionic
MgO	2800	–	Sl sol	Insol	Low in sat'd soln	Ionic
$CoCl_2$	Sublimes	1049	Sol	Insol	High in soln	Ionic
$CoCl_2\ 6H_2O$	86	Dec	Sol	Insol	High in soln	Ionic hydrate, $-H_2O$ at 110°C
$C_{10}H_8$	70	255	Insol	Sol	Zero in melt	Molecular
C_6H_5COOH	122	249	Sl sol	Sol	Low in sat'd soln	Molecular-ionic
$FeCl_3$	282	315	Sol	Insol	High in soln	Molecular-ionic
SnI_4	144	341	Dec	Sol	~ Zero in melt	Molecular
SiO_2	1600	2590	Insol	Insol	Zero in solid	Macromolecular
Fe	1535	3000	Insol	Insol	High in solid	Metallic

Key: Sol = at least 0.1 M; Sl sol = appreciable solubility but <0.1 M; Insol = essentially insoluble; Dec = decomposes.

MOLECULAR SUBSTANCES

All gases and essentially all liquids at room temperature are molecular in nature. If the molecular mass of a substance is over about a hundred, it may be a solid at that temperature. The melting points of molecular substances are usually below 300°C; these substances are relatively volatile, but a good many will decompose before they boil. Most molecular substances do not conduct the electric current either when solid or when molten.

Organic compounds, which contain primarily carbon and hydrogen, often in combination with other nonmetals, are essentially molecular in nature. Since there are a great many organic substances, it is true that most substances are molecular. If an organic compound decomposes on heating, the residue is frequently a black carbonaceous material. Reasonably large numbers of inorganic substances are also molecular; those which are solids at room temperature include some of the binary compounds of elements in Groups IV A, V A, VI A, and VII A.

Molecular substances are frequently soluble in at least a few organic solvents, with the solubility being enhanced if the substance and the solvent are similar in molecular structure.

Some molecular compounds are markedly polar, which tends to increase their solubility in water and other polar solvents. Such substances may ionize appreciably in water, or even in the melt, so that they become conductors of electricity. Often the conductivity is considerably lower than that of an ionic material. Most polar molecular compounds in this category are organic, but a few, including some of the salts of the transition metals, are inorganic.

MACROMOLECULAR SUBSTANCES

Macromolecular substances are all solids at room temperature. They have very high melting points, usually above 1000°C, and low volatility. They are typically very resistant to thermal decomposition. They do not conduct electric current and are often good insulators. They are not soluble in water or any organic solvents. They are frequently chemically inert and may be used as abrasives or refractories.

METALLIC SUBSTANCES

The properties of metals appear to derive mainly from the freedom of movement of their bonding electrons. Metals are good electrical conductors in the solid form, and have characteristic luster and malleability. Most metals are solid at room temperature and have melting points that range from below 0°C to over 2000°C. They are not soluble in water or organic solvents. Some metals are prepared as black powders, which may not appear to be electrical conductors; if such powders are heated, the particles will coalesce to give good electrical conductivity.

EXPERIMENTAL PROCEDURE   WEAR YOUR SAFETY GLASSES WHILE PERFORMING THIS EXPERIMENT

In this experiment you will investigate the properties of several substances with the purpose of determining whether they are ionic, molecular, macromolecular, or metallic. In some cases the classification will be very straightforward. In others you may find that the substance behaves in a way that would not clearly place it in a given category but in some intermediate group.

You may use any tests you wish to on the substances, but use due caution when using materials with which you are not familiar. It is suggested that approximate melting point, solubility in water or organic solvents, electrical conductivity of the solid, liquid, or solution, and tendency to decompose may readily be determined and might aid in classification.

Approximate melting points of substances can be determined rather easily. Substances with low melting points, less than 100°C, for example, will melt readily when warmed gently in a test tube. A test tube heated to about 300°C will impart a yellow-orange color to the Bunsen flame. This color becomes more pronounced between 300° and 550°C, at which temperature the Pyrex tube will begin to soften. When heating samples you should *loosely* stopper the test tube with a cork. Do not breathe any vapors that are given off. Look for water condensing on the cooler portions of the tube and for indications that sublimation is occurring. For the highest temperature studies possible in the lab, heat the sample in a nickel crucible with a strong Bunsen flame; a noticeable red color will appear in the crucible at about 600°C. If this temperature cannot be achieved with a Bunsen burner, use a Meker burner or a gas-air torch. Do not heat samples to 600°C unless their solubility and conductivity properties have been studied at lower temperatures with indecisive results.

Electrical conductivities of your solutions or melts will be measured for you by your laboratory supervisor, who has a portable test meter for that purpose. Distinguish between completely nonconducting, slightly conducting, and highly conducting liquids.

The substances to be studied in the first part of the experiment are on the laboratory tables along with two organic solvents, one polar and one nonpolar. Carry out enough tests on each substance to establish its classification as best you can. Report your observations on each substance, how you would classify it, and your reason for the classification.

When you have completed your tests, report to your laboratory supervisor, who will check your results and issue you two unknowns for characterization.

OBSERVATIONS AND CONCLUSIONS: **Classification of Chemical Substances**

Substance No.	Approximate Melting Point, °C (<100, 100–300, 300–600, >600)	Solubility			Conductivity			Classification and Reason
		H₂O	Nonpolar Organic	Polar Organic	Solid	Melt	Solution in H₂O	
I								
II								
III								
IV								
V								
VI								
Unknown no. ——— ———								

ADVANCE STUDY ASSIGNMENT: Classification of Substances

1. List the properties of a substance which would definitely establish that the material is ionic.

2. If we classify substances as ionic, molecular, macromolecular, or metallic, in which if any categories are all the members

    a. soluble in water?

    b. electrical conductors in the melt?

    c. insoluble in all common solvents?

    d. solids at room temperature?

3. A given substance is a white solid at 25°C. It melts at 250°C without decomposing and the melt does not conduct an electric current. What would be the classification of the substance, based on this information?

4. A white solid melts at 1000°C. The melt does not conduct electricity. Classify the substance as best you can from these properties.

ORGANIC CHEMISTRY

EXPERIMENT

15 • Preparation of Aspirin

One of the simpler organic reactions that one can carry out is the formation of an ester from an acid and an alcohol:

$$
\underset{\text{an acid}}{R\overset{\overset{\displaystyle O}{\|}}{-}C-OH} \; + \; \underset{\text{an alcohol}}{HO-R} \; \rightarrow \; \underset{\text{an ester}}{R\overset{\overset{\displaystyle O}{\|}}{-}C-O-R} \; + \; H_2O \qquad (1)
$$

In the equation, R and R′ are H atoms or organic fragments like CH_3, C_2H_5, or more complex aromatic groups. There are many esters, since there are many organic acids and alcohols, but they all can be formed, in principle at least, by Reaction 1. The driving force for the reaction is in general not very great, so that one ends up with an equilibrium mixture of ester, water, acid, and alcohol.

There are some esters which are solids because of their high molecular mass or other properties. Most of these esters are not soluble in water, so they can be separated from the mixture by crystallization. This experiment deals with an ester of this sort, the substance commonly called aspirin. Aspirin is the active component in headache pills and is one of the most effective, relatively nontoxic, pain killers.

Aspirin can be made by the reaction of the —OH group in the salicylic acid molecule with the carboxyl (—COOH) group in acetic acid:

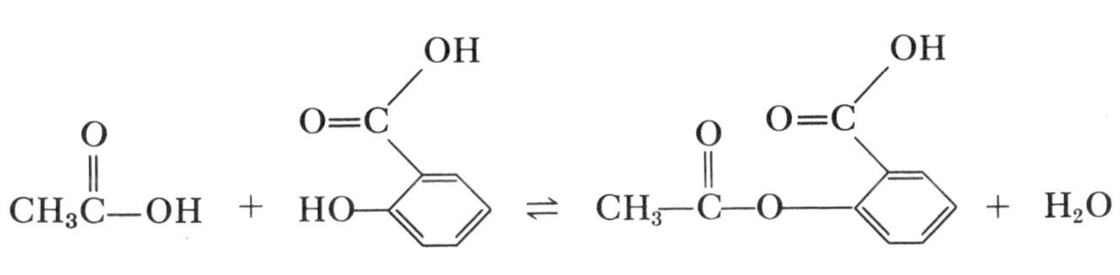

$$(2)$$

acetic acid   salicylic acid   aspirin

105

A better preparative method, which we will use in this experiment, employs acetic anhydride in the reaction instead of acetic acid. The anhydride can be considered to be the product of a reaction in which two acetic acid molecules combine, with the elimination of a molecule of water. The anhydride will react with the water produced in the esterification reaction and will tend to drive the reaction to the right. A catalyst, normally sulfuric or phosphoric acid, is also used to speed up the reaction.

$$\tag{3}$$

acetic anhydride   salicylic acid   aspirin   acetic acid

The aspirin you will prepare in this experiment is relatively impure and should certainly not be taken internally, even if the experiment gives you a bad headache.

EXPERIMENTAL PROCEDURE

WEAR YOUR SAFETY GLASSES WHILE
PERFORMING THIS EXPERIMENT

Weigh a 50 cm³ Erlenmeyer flask on a triple beam or top loading balance and add 2.0 g of salicylic acid. Measure out 5.0 cm³ of acetic anhydride in your graduated cylinder, and pour it into the flask in such a way as to wash any crystals of salicylic acid on the walls down to the bottom. Add 5 drops of 85 per cent phosphoric acid to serve as a catalyst. *Both acetic anhydride and phosphoric acid are reactive chemicals which can give you a bad chemical burn, so use due caution in handling them.* If you get any of either on your hands or clothes, wash thoroughly with soap and water.

Clamp the flask in place in a beaker of water supported on a wire gauze on a ring stand. Heat the water with a Bunsen burner to about 75°C, stirring the liquid in the flask occasionally with a stirring rod. Maintain this temperature for about 15 min, by which time the reaction should be complete. *Cautiously,* add 2 cm³ of water to the flask to decompose any excess acetic anhydride. There will be some hot acetic acid vapor evolved as a result of the decomposition.

When the liquid has stopped giving off vapors, remove the flask from the water bath and add 20 cm³ of water. Let the flask cool for a few minutes in air, during which time crystals of aspirin should begin to form. Put the flask in an ice bath to hasten crystallization and increase the yield of product. If crystals are slow to appear, it may be helpful to scratch the inside of the flask with a stirring rod.

Collect the aspirin by filtering the cold liquid through a Buchner funnel using suction. Turn off the suction and pour about 5 cm³ of ice-cold distilled water over the crystals; after about 15 s turn on the suction to remove the wash liquid along with most of the impurities. Repeat the washing process with another 5 cm³ sample of ice-cold water. Draw air through the funnel for a few minutes to help dry the crystals and then transfer them to a piece of dry filter paper.

While the aspirin is drying, test its solubility properties by taking samples of the solid the size of a pea on your spatula and putting them in separate 1 cm³ samples of each of the following solvents and stirring:

1. Toluene, $CH_3C_6H_5$, nonpolar aromatic
2. Hexane, C_6H_{14}, nonpolar aliphatic
3. Ethyl acetate, $C_2H_5OCOCH_3$, aliphatic ester
4. Ethyl alcohol, C_2H_5OH, polar aliphatic, hydrogen bonding
5. Acetone, CH_3COCH_3, polar aliphatic, nonhydrogen bonding

When the aspirin is dry, weigh the crystals by putting them into a small weighed beaker on the balance and reweighing to 0.1 g. Add 0.5 g to the weight of aspirin obtained to compensate for the amount used in the solubility tests.

Determine the melting point of the aspirin, which is one of the best criteria for its purity. This can be readily done in a small melting point tube, made from 5 mm tubing, as directed by your instructor. Add the crystals to the tube to a depth of about 5 mm, shaking the solid down by tapping the tube on the bench top. Place the tube as shown in Figure 15.1; heat the oil bath *slowly*, especially after the temperature gets over 100°C. As the melting point is approached, the crystals will begin to soften. Report the melting point as the temperature at which the last crystals disappear.

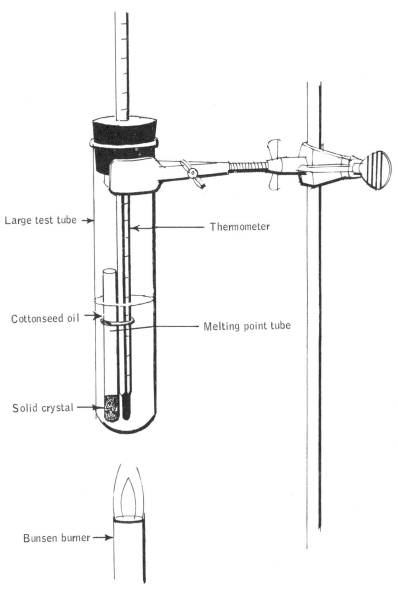

Large test tube →

Thermometer

Cottonseed oil →

Melting point tube

Solid crystal →

Bunsen burner →

Figure 15.1

DATA AND RESULTS: Preparation of Aspirin

Mass of salicylic acid used   _____ g

Volume of acetic anhydride used   _____ ml

Mass of acetic anhydride used
   (density = 1.08 g/cm³)   _____ g

Mass of aspirin obtained   _____ g

Theoretical yield of aspirin   _____ g

Percentage yield of aspirin   _____ %

Melting point of aspirin   _____ °C

Solubility properties of aspirin

    Water   _____   Ethyl acetate _____

    Benzene   _____   Ethyl alcohol _____

    Carbon tetrachloride _____   Acetone   _____

        S = soluble;   I = insoluble;   SS = slightly soluble

Comment on the likely ease in finding a good solvent for an organic solid with the general structural complexity of aspirin.

ADVANCE STUDY ASSIGNMENT: Preparation of Aspirin

1.   Calculate the theoretical yield of aspirin to be obtained in this experiment, starting with 2.0 g of salicylic acid and 5.0 cm³ of acetic anhydride (density = 1.08 g/cm³).

_____ g

2.   If 1.9 g of aspirin were obtained in this experiment, what would be the percentage yield?

_____ %

3.   The name acetic anhydride implies that the compound will react with water to form acetic acid. Write the equation for the reaction.

4.   Identify R and R′ in Equation 1 when the ester, aspirin, is made from salicylic acid and acetic acid.

EXPERIMENT
16 • Preparation of a Synthetic Resin

In this experiment we will prepare and examine the properties of one of the most common polymers, polystyrene, made from styrene, $C_6H_5 - CH = CH_2$. Styrene is relatively easy to polymerize; the plastic made from it is typically quite hard and transparent. In pure polystyrene, the chain is unbranched:

styrene   section of a polystyrene molecule

Polymers containing unbranched chains are thermoplastic, which means that they can be melted and then cast or extruded into various shapes. They also can usually be dissolved in some organic solvents, forming viscous liquids. If a polymer is cross-linked so that its chains are bonded together at regular or random positions, it will usually neither melt nor dissolve readily; such a material is called thermosetting and is usually polymerized in a mold in the shape of the article desired.

Styrene will polymerize to a crystalline solid if you simply heat it. The polymerization reaction itself evolves heat, however, and once the reaction gets started it tends to increase in rate and can get out of control; the simplest commercial process polymerizes styrene this way, and one of the important problems is to provide adequate cooling as the reaction proceeds.

We will polymerize styrene under somewhat different conditions, using an emulsion polymerization, in which the styrene is dispersed into droplets in water. In this process, temperature control is easy, and the polymer is produced in the form of easy-to-handle beads. By carrying out the reaction in the presence of divinylbenzene, which can react at two double-bonded positions, we will make a cross-linked polymer very similar in structure to that of an ion-exchange resin. Divinylbenzene is much like styrene except that there are two ethylene groups rather than one attached to each benzene ring. The compound has three isomers:

paradivinylbenzene   metadivinylbenzene   orthodivinylbenzene

The commercially available divinylbenzene which we use contains mainly the *para* and *meta* isomers, in about equal amounts.

The resin produced will have a structure similar to that indicated below:

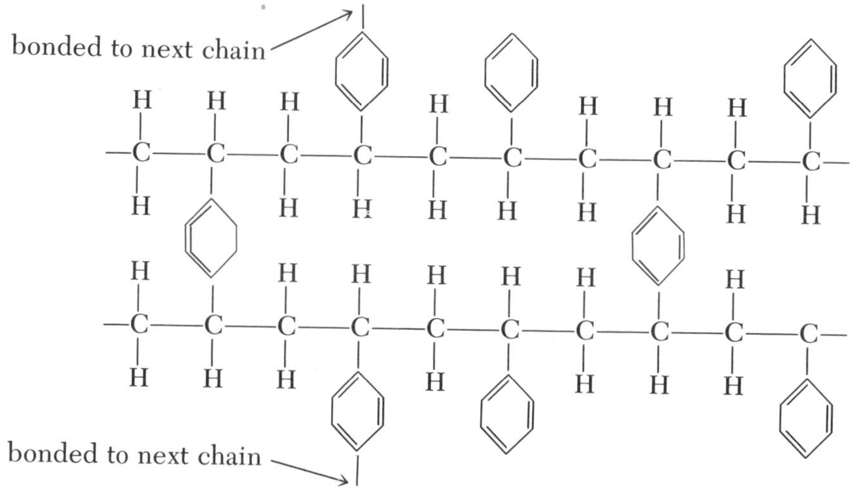

Relatively few divinylbenzene molecules are required for the crosslinking. The material produced is really a copolymer of styrene and divinylbenzene.

After preparing the polymer, we will compare its melting point and solubility properties with those of linear polystyrene.

EXPERIMENTAL PROCEDURE   WEAR YOUR SAFETY GLASSES WHILE PERFORMING THIS EXPERIMENT

Put 100 cm³ water in a 250 cm³ Erlenmeyer flask and heat the flask in a water bath set up as shown in Figure 16.1. When the water in the flask is at about 60°C, slowly add 1.0 g of starch, with stirring; continue stirring and heating until the solution is uniform and the starch completely dispersed.

While the water is heating, measure out 10 cm³ of styrene and 1 cm³ of divinylbenzene into a small beaker. Your instructor will add 100 mg benzoyl peroxide (very reactive!) to the beaker. Stir to dissolve the solid and initiate the reaction; keep the beaker in ice water until the starch solution has been prepared. When the starch solution is ready and at about 80°C, remove it from the water bath and slowly, with swirling, add the styrene solution; stopper the flask *loosely* to minimize vaporization of the styrene. Continue to swirl the liquid for about 20 s to disperse the styrene as small droplets in the starch. Do not shake the flask, since we do not wish to produce a true emulsion, just a dispersion of droplets. Put the flask back in the water bath and heat the mixture in the boiling water for about an hour, during which time it should polymerize completely. Stir every few minutes with your stirring rod to keep the droplets dispersed. If all goes well, the polymer will form as small beads, varying in size from very small up to about 2 mm in diameter.

While the polymerization is proceeding, tear some polystyrene from a coffee cup into small pieces and use it to fill 3 small test tubes. The polymer in the foam is essentially linear polystyrene with few branches and no cross-links. Put 2 cm³ toluene in one of the test tubes and 2 cm³ acetone in the second; shake to get the foam wet with solvent.

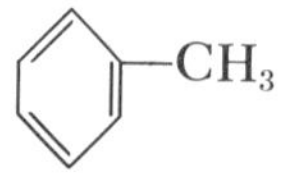

toluene

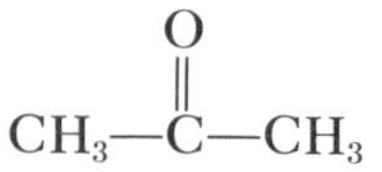

$$CH_3-\overset{\overset{\displaystyle O}{\|}}{C}-CH_3$$

acetone

Heat the third test tube gently in the Bunsen flame, noting whether the polymer melts or decomposes. Estimate the temperature at which the change occurs, but don't try to measure it. Poke the material with your stirring rod to aid in establishing its viscosity. Shake the tubes containing solvent and note whether the polymer has dissolved; if it hasn't, put the tubes in the boiling water in the bath to speed up the solution process. Do not boil the solvent, however. Note the properties of the final mixture, particularly viscosity and clarity of solution.

If you are able to obtain a solution with either solvent, pour a drop of the solution on to some water in a 600 cm³ beaker. Blow gently on the surface to evaporate the solvent completely. Pick up the film with a stirring rod and note its thickness and strength.

When the polymerization reaction is finished, remove the flask from the water bath and pour the slurry into a 600 cm³ beaker half full of distilled water. Stir and then let the beads settle. Decant the liquid and wash twice more to remove any residual starch. Pour the beads out on a paper towel and, when they are dry, weigh them.

Put a few of the beads in small test tubes and test them as before for solubility in toluene and acetone. Try to melt the beads; compare their behavior on heating with that of the polystyrene foam.

CAUTION: In this experiment we use several volatile organic liquids. Avoid breathing their vapors. If convenient, work in a hood or open the windows in the lab. Keep the polymerizing mixture loosely stoppered except when stirring it.

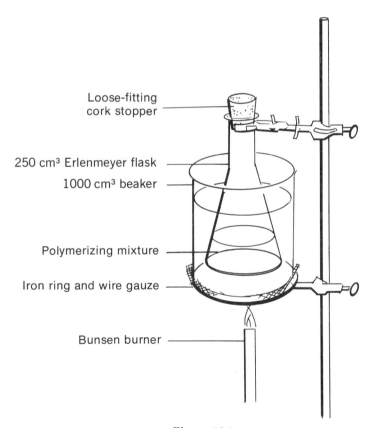

Figure 16.1

DATA AND OBSERVATIONS: Preparation of a Synthetic Resin

Mass of styrene (density = 0.90 g/cm³)   _____ g

Mass of divinylbenzene (density = 0.90 g/cm³)   _____ g

Mass of resin   _____ g

Theoretical yield   _____ g

Percentage yield   _____ %

Properties of polystyrene and prepared resin

	Polystyrene (foam cup)	Prepared resin
Behavior on heating	_____	_____
	_____	_____
Solubility in toluene	_____	_____
Solubility in acetone	_____	_____

How do you explain the difference in solubility of polystyrene in the two solvents?

Comment on the effects of cross-linking on the properties of polystyrene.

ADVANCE STUDY ASSIGNMENT: Preparation of a Synthetic Resin

1. Polypropylene is made by addition polymerization of propylene, $CH_2\!\!=\!\!C\!\!-\!\!H$.
Sketch a section of the polypropylene molecule. $\underset{\displaystyle CH_3}{\overset{\displaystyle |}{}}$

2. How much polypropylene could theoretically be made from 75 g of propylene?

_____ g

3. What per cent by mass of polypropylene is carbon?

_____ %

4. Polystyrene foam such as that used in coffee cups is made by a procedure very analogous to that used in this experiment. Can you suggest how the foamable polymer might be made and how it would be converted to the form of a foam coffee cup?

PROPERTIES OF LIQUIDS AND SOLUTIONS

EXPERIMENT

17 • Vapor Pressure and Heat of Vaporization of Liquids

The vapor pressure of a pure liquid is the total pressure at equilibrium in a container in which only the liquid and its vapor are present. In a container in which the liquid and other gas are both present, the vapor pressure of the liquid is equal to the partial pressure of its vapor in the container. In this experiment you will measure the vapor pressure of a liquid by determining the increase that occurs in the pressure in a closed container filled with air when the liquid is injected into it.

The vapor pressure of a liquid rises rapidly as the temperature is increased and reaches one atmosphere at the normal boiling point of the liquid. Thermodynamic arguments show that the vapor pressure of a liquid depends on temperature according to the equation:

$$\log_{10} VP = -\frac{\Delta H_{vap}}{2.3RT} + C \tag{1}$$

where VP is the vapor pressure, ΔH_{vap} is the amount of heat in joules required to vaporize one mole of the liquid against a constant pressure, R is the gas constant, 8.31 J/mol·K, in the units convenient to this expression, and T is the absolute temperature. You will note that this equation is of the form

$$Y = BX + C \tag{2}$$

where $Y = \log_{10} VP$, $X = 1/T$ and $B = -\Delta H_{vap}/2.3R$. Consequently, if we measure the vapor pressure of a liquid at various temperatures and plot $\log_{10} VP$ vs. $1/T$, we should obtain a straight line. From the slope B of this line, we can calculate the heat of vaporization of the liquid, since $\Delta H_{vap} = -2.3RB$.

In the laboratory you will measure the vapor pressure of an unknown liquid at approximately 0°C, 20°C, and 40°C, as well as its boiling point at atmospheric pressure. Given the three vapor pressures, you will be able to calculate the heat of vaporization of the liquid by making a graph of $\log_{10} VP$ vs. $1/T$. The graph will then be used to predict the boiling point of the liquid, and the value obtained will be compared with that you found experimentally.

121

EXPERIMENTAL PROCEDURE   WEAR YOUR SAFETY GLASSES WHILE
PERFORMING THIS EXPERIMENT

From the stockroom obtain a suction flask, a rubber stopper fitted with a small dropper, and a short length of rubber tubing. Also obtain a sample of an unknown liquid.

1. Assemble the apparatus, using the mercury manometer at your lab bench, as indicated in Figure 17.1. The flask should be dry on the inside. If it is not, rinse it with a little acetone (**flammable**) and blow compressed air into it for a few moments until it is dry. Reassemble the apparatus. Pour some tap water into a beaker and bring it to about 20°C by adding some cold or warm water. Pour this water into the large beaker so that the level of water reaches as far as possible up the neck of the flask. Wait several minutes to ensure that the flask and air inside it are at the temperature of the water. Then remove the stopper from the flask. Pour a small amount of the unknown liquid into a small beaker, and draw about 2 cm³ of the liquid up into the dropper. Blot any excess liquid from the end of the dropper with a paper towel. Press the stopper *firmly* into the flask and connect the hose to the manometer. The mercury levels in the manometer should remain essentially equal.

Immediately squeeze the liquid from the dropper into the flask, where it will vaporize, diffuse, and exert its vapor pressure. This vapor pressure will be equal to the increase in gas pressure at equilibrium in the container. If you do not observe any appreciable (> 10 mm Hg) pressure increase within a minute or two after injecting the liquid, you probably have a leak in your apparatus and should consult your instructor. When the pressure in the flask becomes steady, in about 10 min, read and record the heights of the mercury levels in the manometer to ±1 mm. The difference in height is equal to the vapor pressure of the liquid in mm Hg at the temperature of the water bath. Record that temperature to ±0.2°C.

In this experiment it is essential that (1) the stopper is pressed firmly into the flask, so that the flask plus tubing plus manometer constitute a gas-tight system; (2) no liquid

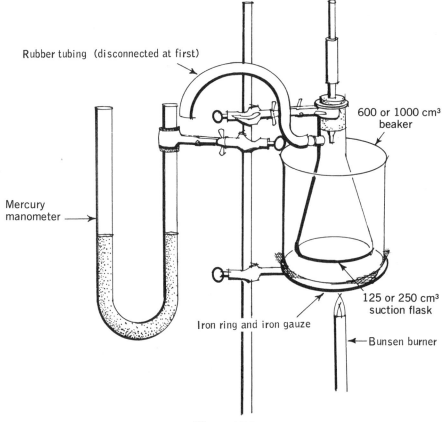

Figure 17.1

falls into the flask until the manometer is connected; (3) the operations of stoppering the flask, connecting the hose to the manometer, and squeezing the liquid into the flask are conducted with dispatch; (4) you take care when connecting the hose to the manometer, since mercury has a poisonous vapor and is not easily cleaned up when spilled.

Remove the flask from the system. Dry it with compressed air, and dry the dropper by squeezing it several times in air.

2. Starting again at 1, carry out the same experiment at about 0°C. Use an ice-water bath for cooling the flask. Measure the latter temperature, rather than assuming it to be 0°C.

3. Repeat the experiment once again, this time holding the water bath at 40°C by judicious heating with a bunsen burner. Between each run, the flask and dropper must be thoroughly dried, and the precautions noted carefully observed.

4. In the last part of the experiment you will measure the boiling point of the liquid at the barometric pressure in the laboratory. This is done by pouring the remaining sample of unknown into a large test tube. Determine the boiling point of the liquid following the procedure which is indicated in Figure 17.2. The thermometer bulb should

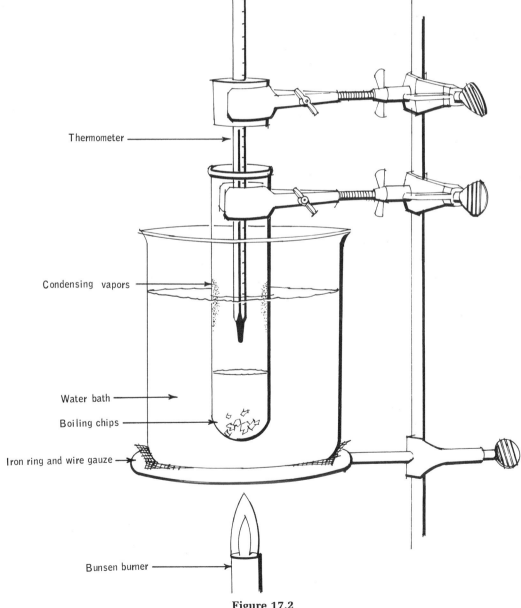

Figure 17.2

be just above the liquid surface. Heat the water bath until the liquid in the tube boils gently, with its vapor condensing *at least 5 cm below* the top of the test tube. Boiling chips may help in keeping the liquid boiling smoothly. The steady temperature obtained under these conditions is the boiling point of the liquid. The liquids used may be flammable, and you should not inhale their vapors unnecessarily so *do not* heat the water bath so strongly that condensation of the vapors from the liquid occurs only at the top of the test tube.

DATA AND CALCULATIONS: Vapor Pressure and Heat of Vaporization of Liquids

Temperature, t, in °C	Heights of manometer mercury levels in mm	Vapor pressure mm Hg	VP (kPa) (1 mm Hg = 0.1333 kPa)
1. _____	_____	_____	_____ _____
2. _____	_____	_____	_____ _____
3. _____	_____	_____	_____ _____

Boiling point _____°C

Barometric pressure _____kPa

Using the relation (1) between vapor pressure and temperature, we will calculate the molar heat of vaporization of your liquid and its boiling point from the vapor pressure data obtained. It would be useful to first make the calculations indicated in the following table:

Approximate temperature °C	t, actual temperature °C	T, temperature K	$1/T$	Vapor pressure, VP, in kPa	$\log_{10} VP$
0	_____	_____	_____	_____	_____
20	_____	_____	_____	_____	_____
40	_____	_____	_____	_____	_____

On the graph paper provided make a graph of $\log_{10} VP$ vs. $1/T$. Let $\log_{10} VP$ be the ordinate and plot $1/T$ on the abscissa. Since $\log_{10} VP$ is, by (1), a linear function of $1/T$, the line obtained should be nearly straight. Find the slope of the line, $\Delta\log_{10} VP/\Delta(1/T)$.

The slope of the line is equal to $-\Delta H_{\text{vaporization}}/2.3R$. Given that $R = 8.31$ J/mol · K, calculate the molar heat of vaporization, ΔH_{vap}, of your liquid.

$$\text{Slope} = \frac{\Delta \log_{10} VP}{\Delta 1/T} = \underline{\hspace{2cm}} = \frac{-\Delta H_{\text{vap}}}{2.3R} \quad , \qquad \Delta H_{\text{vap}} = \underline{\hspace{2cm}} \text{ J/mol}$$

Continued on following page   **125**

   From the graph it is also possible to find the temperature at which your liquid will have any given vapor pressure. Recalling that a liquid will boil in an open container at the *temperature* at which its *vapor pressure* is equal to the *atmospheric pressure*, predict the boiling point of the liquid at the barometric pressure in the laboratory.

$P_{barometric}$ _____ kPa

$\log_{10} P_{barometric}$ _____

$1/T$ at this pressure _____ K^{-1} (from graph)

T at this pressure _____ K

t at this pressure _____ °C (boiling point predicted)

Boiling point observed experimentally _____ °C

Unknown no. _____

   Comment on the validity of the equation:

$$\log_{10} VP = \frac{-\Delta H_{vap}}{2.3RT} + C$$

Name_____   Section_____

VAPOR PRESSURE AND HEAT OF
VAPORIZATION OF LIQUIDS
(Data and Calculations)

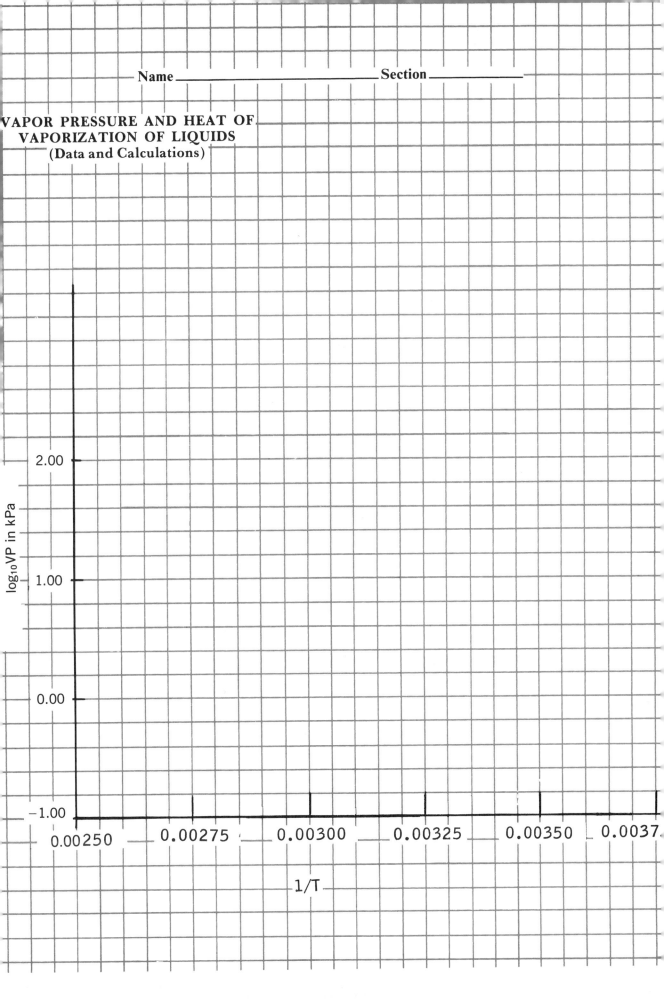

ADVANCE STUDY ASSIGNMENT: Vapor Pressure of Liquids

1. When a little liquid carbon tetrachloride is introduced at 40°C into a closed container in which the gas pressure is 99.0 kPa, the pressure rises to 132.3 kPa, and the addition of more carbon tetrachloride does not increase the pressure. What is the vapor pressure of CCl_4 at 40°C?

————————— kPa

2. In a vapor pressure experiment on benzene, C_6H_6, the following data were obtained:

t °C	0	25	60
VP_{kPa}	3.6	12.6	51.7

a. Prepare a table of $\log_{10} VP$ vs. $1/T$ (T is in K)

$\log_{10} VP$	$1/T$
————————	————————
————————	————————
————————	————————

b. Using the graph paper on the reverse side of this sheet, plot $\log_{10} VP$ vs. $1/T$.
c. Draw a straight line through the three points.
d. Determine the slope of the line.

Slope = —————————

e. Using equations (1) and (2), calculate ΔH_{vap}.

ΔH_{vap} = ————————— J/mol

3. Benzene will boil in a beaker when its vapor pressure exceeds the barometric pressure at its surface. Using the graph made in Problem 2, find the temperature at which the vapor pressure of benzene equals 101.3 kPa. This is the normal boiling point of benzene.

BP = ————————— °C

BP observed = ———80.1——— °C

VAPOR PRESSURE OF LIQUIDS
(Advance Study Assignments)

18 • Molecular Mass Determination by Depression of the Freezing Point

In the preceding experiment you observed the change of vapor pressure of a liquid as a function of temperature. If a nonvolatile solid compound (the solute) is dissolved in a liquid, the vapor pressure of the liquid solvent is lowered. This decrease in the vapor pressure of the solvent results in other easily observable physical changes; the boiling point of the solution is higher than that of the pure solvent and the freezing point is lower.

Many years ago chemists observed that at low solute concentrations the changes in the boiling point, the freezing point, and the vapor pressure of a solution are all proportional to the amount of solute that is dissolved in the solvent. These three properties are collectively known as colligative properties of solutions. The colligative properties of a solution depend only on the number of solute particles present in a given amount of solvent and not on the kind of particles dissolved.

When working with boiling point elevations or freezing point depressions of solutions, it is convenient to express the solute concentration in terms of its molality m defined by the relation:

$$\text{molality of } A = m_A = \frac{\text{no. of moles } A \text{ dissolved}}{\text{no. of kilograms of solvent}}$$

For this unit of concentration, the boiling point elevation, $T_b - T_b^\circ$ or ΔT_b, and the freezing point depression, $T_f^\circ - T_f$ or ΔT_f, in °C at low concentrations are given by the equations:

$$\Delta T_b = k_b m \qquad \Delta T_f = k_f m \qquad (1)$$

where k_b and k_f are characteristic of the solvent used. For water, $k_b = 0.52$ and $k_f = 1.86$. For benzene, $k_b = 2.53$ and $k_f = 5.10$.

One of the main uses of the colligative properties of solutions is in connection with the determination of the molecular masses of unknown substances. If we dissolve a known amount of solute in a given amount of solvent and measure ΔT_b or ΔT_f of the solution produced, and if we know the appropriate k for the solvent, we can find the molality and hence the molecular mass of the solute. In the case of the freezing point depression, the relation would be:

$$\Delta T_f = k_f m = k_f \times \frac{\text{no. moles solute}}{\text{no. kilograms solvent}} = k_f \times \frac{\dfrac{\text{no. g solute}}{\text{GMM solute}}}{\text{no. kilograms solvent}} \qquad (2)$$

In this experiment you will be asked to estimate the molecular mass of an unknown solute, using this equation. The solvent used will be paradichlorobenzene, which has a convenient melting point and a relatively large value for k_f, 7.10. The freezing points will be obtained by studying the rate at which liquid paradichlorobenzene and some of its solutions containing the unknown cool in air. Temperature-time graphs, called cooling curves, reveal the freezing points very well, since the rate at which a liquid cools is typically quite different from that of a liquid-solid equilibrium mixture.

EXPERIMENTAL PROCEDURE

A. Determination of the Freezing Point of Paradichlorobenzene. From the stockroom obtain a stopper fitted with a sensitive thermometer and a glass stirrer, a large test tube, and a sample of solid unknown. **Remember that the thermometer is both fragile and expensive, so handle it with due care.** Weigh the test tube on a top loading or triple beam balance to 0.01g. Add about 30 g of paradichlorobenzene, PDB, to the test tube and weigh again to the same precision.

Fill your 600 cm³ beaker almost full of hot water from the faucet. Support the beaker on an iron ring and wire gauze on a ring stand and heat the water with a Bunsen burner. Clamp the test tube to the ring stand and immerse the tube in the water as far as is convenient.

Heat the water to about 70 to 75°C (use your ordinary thermometer to follow the bath temperature), at which point most of the PDB will melt. Insert the stopper-thermometer-stirrer assembly in the test tube, adjusting the level of the thermometer so that the bulb is 1 cm above the bottom of the tube, well down into the melt. When the PDB is at about 65°C, stop heating. Stir, to dissolve any remaining solid PDB. Carefully lower the iron ring and water bath and put the beaker of hot water on the lab bench well away from the test tube. Dry the outside of the test tube with a towel.

Record the temperature of the paradichlorobenzene as it cools in the air. Stir the liquid slowly but continuously to avoid supercooling. Start readings at about 60°C and note the temperature every 30 s for eight min or until the solution has solidified to the point that you are no longer able to stir it. Near the melting point you will begin to observe crystals of PDB in the liquid, and these will increase in amount as cooling proceeds. Note the temperature at which the first solid PDB appears. In Figure 18.1 we have shown graphically how the temperature of pure PDB will typically vary with time in this experiment.

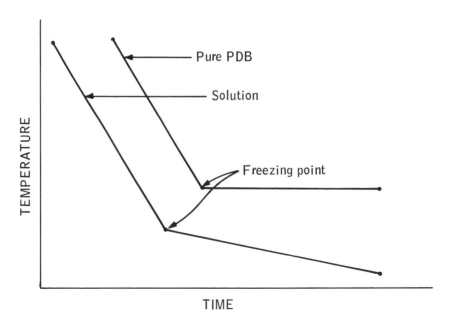

Figure 18.1

B. Determination of the Molecular Mass of an Unknown Compound. Weigh your unknown in its container to 0.01 g. Pour about half of the sample (about 2 g) into your test tube of PDB and reweigh the container.

Heat the test tube in the water bath until the PDB is again melted and the solid unknown is dissolved. When the melt has reached about 65°C, remove the bath, dry the test tube, and let it cool as before. Start readings at about 60°C and continue to take readings, with stirring, for eight minutes.

The dependence of temperature on time with the solution will be similar to that observed for pure PDB, except that the first crystals will appear at a lower temperature, and the temperature of the solid-solution system will gradually fall as cooling proceeds. There may be some supercooling, as evidenced by a rise in temperature shortly after the first appearance of PDB crystals. In this case the freezing point of the solution is best taken as the temperature at which the two lines on the temperature-time graph intersect (Figure 18.1).

Repeat the experiment with this same solution to check for reproducibility of your data.

Add the rest of your unknown (about 2 g) to the PDB solution and again weigh the container. Melt the PDB as before, heating it to about 65°C before removing the water bath. Repeat the entire procedure described above, again checking for reproducibility of data.

DATA AND CALCULATIONS: **Molecular Mass Determination by Freezing Point Depression**

Mass of large test tube   _____ g

Mass of test tube plus about 30 g of paradichlorobenzene   _____ g

Mass of container plus unknown   _____ g

Mass of container less Sample I   _____ g

Mass of container less Sample II   _____ g

Time-Temperature Readings

Temperature

Time (minutes)	Paradichloro-benzene	Solution I 1st run	Solution I 2nd run	Solution II 1st run	Solution II 2nd run
0	_____	_____	_____	_____	_____
$1/2$	_____	_____	_____	_____	_____
1	_____	_____	_____	_____	_____
$1 1/2$	_____	_____	_____	_____	_____
2	_____	_____	_____	_____	_____
$2 1/2$	_____	_____	_____	_____	_____
3	_____	_____	_____	_____	_____
$3 1/2$	_____	_____	_____	_____	_____
4	_____	_____	_____	_____	_____
$4 1/2$	_____	_____	_____	_____	_____
5	_____	_____	_____	_____	_____
$5 1/2$	_____	_____	_____	_____	_____

Continued on following page   **135**

6	_____	_____	_____	_____	_____
6½	_____	_____	_____	_____	_____
7	_____	_____	_____	_____	_____
7½	_____	_____	_____	_____	_____
8	_____	_____	_____	_____	_____

Approximate temperature at which
first solid appeared

_____ _____ _____ _____ _____

Estimation of Freezing Point

On the graph paper provided, plot your temperature vs. time readings for pure paradichlorobenzene and each run of the two solutions. To avoid overlapping graphs, add four minutes to all observed times in making the graph of the cooling curve for pure paradichlorobenzene. Add three minutes to all times for Solution I, run 1; two minutes for Solution I, run 2; one minute for Solution II, run 1; and use times as observed for Solution II, run 2. The freezing point of the liquid may, in each case, be taken to be the point of intersection of the approximately straight line portions of the cooling curve.

Calculation of Molecular Mass

	Solution I	Solution II
Mass of unknown used (total amount in solution)	_____ g	_____ g
Mass of paradichlorobenzene used	_____ g	_____ g
Freezing point of pure paradichlorobenzene	_____ °C	_____ °C
Freezing point of solution (average values of runs 1 and 2)	_____ °C	_____ °C
Freezing point depression (total depression)	_____ °C	_____ °C
Total molal concentration of unknown solution (from Equation 2) ($k_f = 7.10$)	_____	_____

Continued on following page

Gram molecular mass of unknown
(from Equation 2)   _____   _____

Average MM   _____

Unknown no.   _____

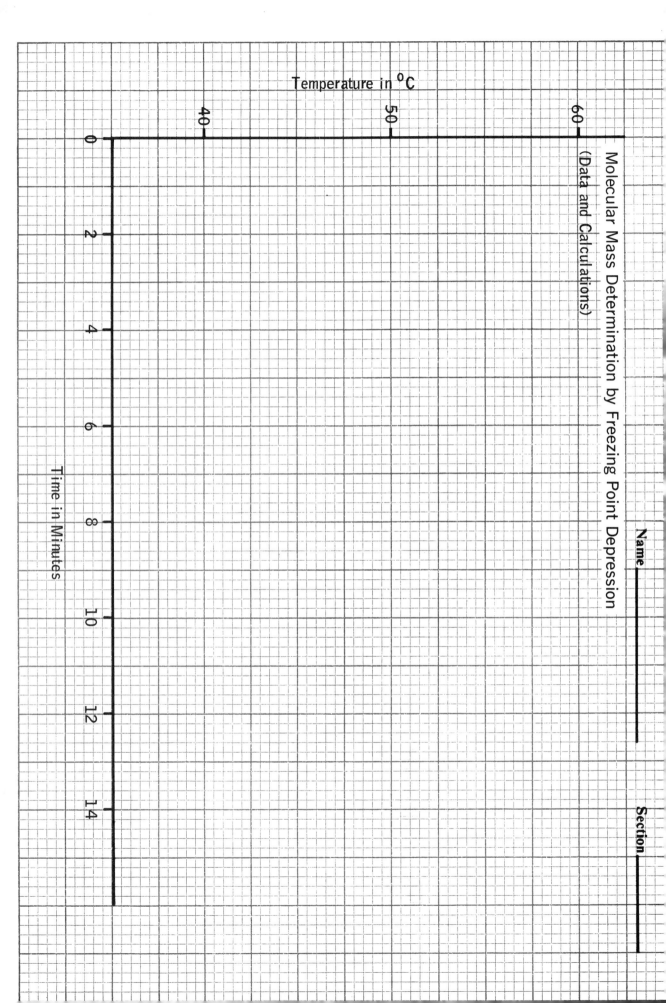

Molecular Mass Determination by Freezing Point Depression

(Data and Calculations)

Name

Section

Temperature in °C

40

50

60

Time in Minutes

0

2

4

6

8

10

12

14

**ADVANCE STUDY ASSIGNMENT: Determination of Molecular Mass by
Freezing Point Depression**

1. A student determines the freezing point of a solution of 1.96 g of naphthalene in
25.64 g of paradichlorobenzene. He obtains the following temperature-time readings.

Time (min)	0	1/2	1	1 1/2	2	2 1/2
Temperature (°C)	59.7	58.0	56.5	54.8	53.4	52.1

Time (min)	3	3 1/2	4	4 1/2	5	5 1/2
Temperature (°C)	50.9	49.5	48.4	48.6	48.7	48.6

Time (min)	6	6 1/2	7	7 1/2	8
Temperature (°C)	48.5	48.4	48.3	48.2	48.1

(a) Plot these data on the graph paper provided. Estimate the freezing point of the
solution to ±0.1°C by determining the intersection of the straight line portions of the
cooling curves.

_____°C

(b) Taking k_f for paradichlorobenzene to be 7.10, calculate the molecular mass
of the solid. Assume 53.0°C to be the freezing point of pure paradichlorobenzene.

MM = _____

2. Precise freezing point determinations show that Equation 2 is really valid only for
dilute solutions and that at high concentrations, the molecular masses obtained are
appreciably in error. By calculating apparent molecular masses at various concentrations by Equation 2 and extrapolating those values to zero concentration, one can obtain
the vest value possible for the molecular mass of a given solute.

Continued on following page   **139**

Sucrose solutions in water, in very precise experiments, are found to have the following freezing points:

g sucrose/kg H$_2$O	T_f, °C	m(effective)	MM (apparent)
100	−0.56	_____	_____
200	−1.15	_____	_____
400	−2.42	_____	_____
600	−4.05	_____	_____

Given that k_f for water is 1.86, use Equation 2 to calculate the effective molality m and the apparent molecular mass of sucrose in each solution. Make a graph of MM apparent vs. g sucrose present, and extrapolate the line to g sucrose equals zero to find the best value of the molecular weight of sucrose. (We will not use this procedure in the experiment you perform, since experimental errors are too high. Under such conditions, averaged results are probably best.)

Name _____ Section _____

Temperature in °C

Time in minutes

Apparent molecular mass

g sucrose/ kg H$_2$0

SECTION EIGHT
CHEMICAL EQUILIBRIUM

EXPERIMENT

19 • Determination of the Equilibrium Constant for a Chemical Reaction

Every chemical reaction has a characteristic condition of equilibrium at a given temperature. If two reactants are mixed, they will tend to react to form products until a state is reached where the amounts of reactants and products no longer change. Under such conditions the reactants and products are in chemical equilibrium and will remain so until the system is altered in some way. Associated with the equilibrium state there is a number called the equilibrium constant, K_c, which expresses the necessary condition on the concentrations of reactants and products for the reaction. For the reaction

$$aA + bB \rightleftharpoons cC + dD \tag{1}$$

the equilibrium condition* is that

$$\frac{[C]^c[D]^d}{[A]^a[B]^b} = K_c \tag{2}$$

where A, B, C, and D are solutes in solution or are gases, and the bracketed expressions are their concentrations in moles per cubic decimetre. The equilibrium constant K_c,

---

*The expression for the equilibrium constant K_c is fundamental and will appear in several of its many possible forms in many of the experiments in the latter part of this manual.

Strictly speaking, the expression for K_c should be set up in terms of the activities of the species A, B, C, and D, rather than their concentrations. The activity of a species may be considered to be its effective concentration in a system, rather than its actual concentration. At low concentrations the activities of solute and gaseous species and their concentrations are nearly equal. In most applications of the equilibrium concept, whether they be in homogeneous systems (as in this experiment) or in heterogeneous systems, in which solubilities are of concern, concentrations of species are used rather than activities in the expression for K_c. This practice will be followed in all experiments involving chemical equilibria that appear in this manual.

In the experiments on equilibrium systems an attempt will be made to illustrate various aspects of equilibrium theory. In order to make the experiments tractable, it will be necessary in some cases to make simplifying assumptions about the system being studied. In addition to assuming that concentrations may be employed instead of activities, we will in general limit the number of equilibria to be considered in any given system. Since there are many equilibria that are simultaneously satisfied in some systems, this limitation may produce results that are not in complete accord with data in the literature. In some instances, we admit, we have sacrificed a complete treatment of a system in order to more clearly illustrate a valid principle.

143

will have a fixed value for the reaction at any given temperature. If A, B, C, and D are mixed in arbitrary amounts in a container they will tend to react until their concentrations satisfy Equation 2. Depending on the magnitude of K_c and the amounts of species used initially, Reaction 1 will proceed to the right or to the left until equilibrium is attained.

In this experiment we will study the equilibrium properties of the reaction between iron(III) ion and thiocyanic acid:

$$Fe^{3+}(aq) + HSCN(aq) \rightleftharpoons FeSCN^{2+}(aq) + H^+(aq) \tag{3}$$

When solutions containing Fe^{3+} ion and thiocyanic acid are mixed, Reaction 3 occurs to some extent, forming the $FeSCN^{2+}$ complex ion, which has a deep red color, and H^+ ion. As a result of the reaction, the equilibrium amounts of Fe^{3+} and HSCN will be less than they would have been if no reaction occurred; for every mole of $FeSCN^{2+}$ that is formed, one mole Fe^{3+} and one mole of HSCN will react. According to the general law, K_c for Reaction 3 takes the form

$$\frac{[FeSCN^{2+}][H^+]}{[Fe^{3+}][HSCN]} = K_c \tag{4}$$

Our purpose in the experiment will be to evaluate K_c for the reaction by determining the equilibrium concentrations of the four species in Equation 4 in several solutions made up in different ways. The equilibrium constant K_c for the reaction has a convenient magnitude and the color of the $FeSCN^{2+}$ ion makes for an easy analysis of the equilibrium mixture.

The solutions will be prepared by mixing solutions containing known concentrations of iron(III) nitrate and thiocyanic acid. The color of the $FeSCN^{2+}$ ion formed will allow us to determine its equilibrium concentration. Knowing the initial composition of the solution and the equilibrium concentration of $FeSCN^{2+}$, we can calculate the equilibrium concentrations of the rest of the pertinent species and then calculate K_c.

Since the calculations that are necessary to find K_c may not be apparent, let us consider a specific example. Assume that we prepare our solution by mixing $10.0 \text{ cm}^3 \ 2.00 \times 10^{-3} \ M \ Fe(NO_3)_3$ with $10.0 \text{ cm}^3 \ 2.00 \times 10^{-3} \ M$ HSCN under conditions which keep $[H^+]$ equal to $0.50 \ M$. The Fe^{3+} in the iron(III) nitrate reacts with the HSCN to produce some red $FeSCN^{2+}$ complex ion. By a method of analysis described below it is found that $[FeSCN^{2+}]$ at equilibrium is about $1.50 \times 10^{-4} \ M$.

To find K_c for the reaction from these data it is convenient first to determine how many moles of reactant species were initially present, before the reaction occurred. By the definition of the molarity M of a species A,

$$M_A = \frac{\text{no. moles } A}{\text{no. dm}^3 \text{ solution}} \quad \text{or} \quad \text{no. moles } A = M_A V$$

where V is the volume of solution in cubic decimetres.

$$\text{Initial no. moles } Fe^{3+} = M_{Fe^{3+}} V_{Fe(NO_3)_3} = 2.00 \times 10^{-3} \times \frac{10.0 \text{ cm}^3}{1000 \text{ cm}^3/\text{dm}^3} = 20.0 \times 10^{-6}$$

$$\text{Initial no. moles HSCN} = M_{HSCN} V_{HSCN} = 2.00 \times 10^{-3} \times \frac{10.0 \text{ cm}^3}{1000 \text{ cm}^3/\text{dm}^3} = 20.0 \times 10^{-6}$$

The number of moles $FeSCN^{2+}$ present at equilibrium is found from the analysis and the volume of the solution:

$$\text{Equilibrium no. moles } FeSCN^{2+} = M_{FeSCN^{2+}} V = 1.50 \times 10^{-4} \times \frac{20.0 \text{ cm}^3}{1000 \text{ cm}^3/\text{dm}^3} = 3.00 \times 10^{-6}$$

since the volume of the solution containing the complex ion is 20.0 cm³. The $FeSCN^{2+}$ ion is produced by Reaction 3:

$$Fe^{3+}(aq) + HSCN(aq) \rightarrow FeSCN^{2+}(aq) + H^+(aq)$$

Therefore, for every mole of $FeSCN^{2+}$ present in the equilibrium mixture, one mole Fe^{3+} and one mole HSCN are reacted. We can see then that

Equilibrium no. moles Fe^{3+} = no. moles Fe^{3+} initially present − no. moles $FeSCN^{2+}$ at equilibrium

$$= 20.0 \times 10^{-6} - 3.00 \times 10^{-6} = 17.0 \times 10^{-6}$$

Similarly,

Equilibrium no. moles HSCN = no. moles HSCN initially present − no. moles $FeSCN^{2+}$ at equilibrium

$$= 20.0 \times 10^{-6} - 3.00 \times 10^{-6} = 17.0 \times 10^{-6}$$

Knowing the no. moles Fe^{3+} and HSCN present in the equilibrium mixture, and the volume of the mixture, we can easily find the concentrations of those two species:

$$[Fe^{3+}] = \frac{\text{no. moles } Fe^{3+}}{\text{no. dm}^3 \text{ solution}} = \frac{17.0 \times 10^{-6}}{20.0 \times 10^{-3}} = 8.5 \times 10^{-4} \ M$$

$$[HSCN] = \frac{\text{no. moles HSCN}}{\text{no. dm}^3 \text{ solution}} = \frac{17.0 \times 10^{-6}}{20.0 \times 10^{-3}} = 8.5 \times 10^{-4} \ M$$

We now can substitute into Equation 4 to find the equilibrium constant for this reaction:

$$K_c = \frac{[FeSCN^{2+}][H^+]}{[Fe^{3+}][HSCN]} = \frac{1.50 \times 10^{-4} \times 0.50}{8.5 \times 10^{-4} \times 8.5 \times 10^{-4}} = 104$$

(The data in this calculation correspond to a different temperature than the one at which you will be working, so the actual value of K_c you will obtain will not be the one found above.)

Two methods of analysis can be easily used to determine $[FeSCN^{2+}]$ in the equilibrium mixtures. The more precise method makes use of a spectrophotometer, which measures the amount of light absorbed by the red complex at 447 nm, the wavelength at which the complex most strongly absorbs. The absorbance A of the complex is proportional to its concentration M and can be measured directly on the spectrophotometer:

$$A = kM$$

Your instructor will show you how to operate the spectrophotometer if these are available to your laboratory and will provide you with a calibration curve from which you can find $[FeSCN^{2+}]$ once you have determined the absorbance of your solutions.

In the other method for analysis a solution of known concentration of $FeSCN^{2+}$ is prepared. By determining the depth of this standard solution, which matches in color intensity the unknown solution of known depth, one can calculate $[FeSCN^{2+}]$.

In all the solutions to be mixed we will maintain $[H^+]$ at a value of 0.5 M. This large amount of H^+ ion will ensure that essentially all the HSCN present remains undissociated and that no Fe^{3+} reacts to form other brown colored species such as $FeOH^{2+}$. Since H^+ will be present in large excess as compared to the other reactants, its concentration will not be appreciably affected by the reaction which occurs and can be assumed to remain constant and equal to 0.5 M.

EXPERIMENTAL PROCEDURE

Label five test tubes 1 through 5. Pipet 5.0 cm^3 of 2.00×10^{-3} M Fe(NO$_3$)$_3$ in 0.50 M HNO$_3$ into each test tube. Into each of the test tubes labeled 1 to 5, pipet the corresponding number of cm^3 (1 to 5) of 2.00×10^{-3} M HSCN in 0.50 M HNO$_3$. Add enough 0.50 M HNO$_3$ to each test tube to make the total volume equal to 10.0 cm^3. Mix each solution thoroughly with a glass stirring rod. Be sure to dry the stirring rod after mixing each solution.

Method I. Analysis by Spectrophotometric Measurement.   Place a portion of each solution in a spectrophotometer cell as shown by your instructor and measure the absorbance of the solution at 447 nm. Determine the concentration of FeSCN^{2+} from the calibration curve provided for each instrument. Record these values on the data page.

Method II. Analysis by Comparison with a Standard.   Prepare a solution of known [FeSCN^{2+}] by pipetting 5.0 cm^3 0.200 M Fe(NO$_3$)$_3$ in 0.50 M HNO$_3$ into a test tube and adding 1.0 cm^3 0.002 M HSCN in 0.50 M HNO$_3$. Dilute to 20 cm^3 with 0.50 M HNO$_3$. Mix the solution thoroughly with a stirring rod.

Since in this solution [Fe^{3+}] $\gg$ [HSCN], Reaction 3 is driven strongly to the right. You can assume without serious error that essentially all the HSCN added is converted to FeSCN^{2+}. Assuming that this is the case, calculate [FeSCN^{2+}] in the standard solution and record the value on the data page.

The [FeSCN^{2+}] in the unknown solutions in test tubes 1 to 5 can be found by comparing the intensity of the red color in those solutions with that in the standard solution. This can be done by placing the test tube containing Solution 1 side by side with the test tube containing the standard. Look down both test tubes toward a well-illuminated piece of white paper on the laboratory bench.

Pour out the standard solution into a dry clean beaker until the color intensity you see down the tube containing the standard matches that which you see when looking down the tube containing the unknown. When the colors match, the following relation is valid:

[FeSCN^{2+}]$_{\text{unknown}}$ × depth of unknown solution

$$= [\text{FeSCN}^{2+}]_{\text{standard}} \times \text{depth of standard solution} \qquad (6)$$

Measure the depths of the matching solutions with a rule and record them. Repeat the measurement for solutions 2 through 5, recording the depth of each unknown and that of the standard solution which matches it in intensity.

DATA: Determination of the Equilibrium Constant for a Chemical Reaction

SOLUTION	2.00×10^{-3} M Fe(NO$_3$)$_3$	2.00×10^{-3} M HSCN	0.5 M HNO$_3$	METHOD I absorbance	METHOD II depth standard	METHOD II depth unknown	[FeSCN^{2+}]
	CUBIC CENTIMETRES						
1	_____	_____	_____	_____	_____	_____	_____ $\times 10^{-4}$
2	_____	_____	_____	_____	_____	_____	_____ $\times 10^{-4}$
3	_____	_____	_____	_____	_____	_____	_____ $\times 10^{-4}$
4	_____	_____	_____	_____	_____	_____	_____ $\times 10^{-4}$
5	_____	_____	_____	_____	_____	_____	_____ $\times 10^{-4}$

If Method II was used: [FeSCN^{2+}]$_{standard}$ _____ $\times 10^{-4}$ M; [FeSCN^{2+}] in solutions 1 to 5 calculated by Relation 6.

CALCULATIONS

A. Calculation of K_c assuming the reaction: Fe^{3+}(aq) + HSCN(aq) $\rightleftharpoons$ FeSCN^{2+}(aq) + H$^+$(aq)

This calculation is most readily carried out by completing the table on p. 148 as follows:

1. Knowing the initial conc. and volume of Fe^{3+} and HSCN, calculate the initial no. of moles of these species (express as a number $\times 10^{-6}$).

2. Record [FeSCN^{2+}] from above. Calculate the no. of moles of FeSCN^{2+} in your 10 cm^3 of solution. Again, express as a number $\times 10^{-6}$.

3. Realizing that one mole of FeSCN^{2+} is formed at the expense of one mole of Fe^{3+} and one mole of HSCN, calculate the number of moles of Fe^{3+} and HSCN at equilibrium.

4. Knowing the no. of moles of Fe^{3+} and HSCN at equilibrium and the volume (10 cm^3), calculate [Fe^{3+}] and [HSCN]. Express as a number $\times 10^{-4}$.

5. Use Equation 4 to calculate K_c.

Continued on following page

SOLUTION	INITIAL NO. MOLES		EQUILIBRIUM NO. MOLES			EQUILIBRIUM CONCENTRATIONS				
	Fe^{3+} $\times 10^{-6}$	HSCN $\times 10^{-6}$	Fe^{3+} $\times 10^{-6}$	HSCN $\times 10^{-6}$	$FeSCN^{2+}$ $\times 10^{-6}$	Fe^{3+} $\times 10^{-4}$	[HSCN] $\times 10^{-4}$	$[FeSCN^{2+}]$ $\times 10^{-4}$	$[H^+]$	K_c
1	___	___	___	___	___	___	___	___	0.50	___
2	___	___	___	___	___	___	___	___	0.50	___
3	___	___	___	___	___	___	___	___	0.50	___
4	___	___	___	___	___	___	___	___	0.50	___
5	___	___	___	___	___	___	___	___	0.50	___

B. In calculating K_c in Part **A**, we assumed that the formula of the complex ion is $FeSCN^{2+}$. It is by no means obvious that this is the case, and one might have assumed, for instance, that $Fe(SCN)_2^+$ was the species formed. The reaction would then be:

$$Fe^{3+}(aq) + 2\ HSCN(aq) \rightleftharpoons Fe(SCN)_2^+(aq) + 2\ H^+(aq) \tag{7}$$

If one analyzed the equilibrium system we have studied, assuming that Reaction 7 occurs rather than Reaction 3, we would presumably obtain nonconstant values of K_c. Using the same kind of procedure as in Part **A**, calculate K_c for solutions 1, 3, and 5 on the basis that $Fe(SCN)_2^+$ is the formula of the complex ion formed by reaction between Fe^{3+} and HSCN. Because of the procedure used for calibrating the system by Method I or Method II, $[Fe(SCN)_2^+]$ will equal *one-half* the $[FeSCN^{2+}]$ obtained for each solution in Part **A**. Note that *two* moles HSCN are needed to form *one* mole $Fe(SCN)_2^+$; this changes not only the relative numbers of moles from the previous case but also the expression for K_c.

SOLUTION	INITIAL NO. MOLES		EQUILIBRIUM NO. MOLES			EQUILIBRIUM CONCENTRATIONS				
	Fe^{3+}	HSCN	$Fe(SCN)_2^+$	Fe^{3+}	HSCN	$[Fe(SCN)_2^+]$	$[Fe^{3+}]$	[HSCN]	$[H^+]$	K_c
1	___	___	___	___	___	___	___	___	___	___
3	___	___	___	___	___	___	___	___	___	___
5	___	___	___	___	___	___	___	___	___	___

On the basis of the results of Part **A**, what can you conclude about the validity of the equilibrium concept, as exemplified by Equation 4? What can you conclude about the formula of the iron(III) thiocyanate complex ion?

ADVANCE STUDY ASSIGNMENT: Determination of the Equilibrium
Constant for a Chemical Reaction

1.   When Fe^{3+} and HSCN react to an equilibrium with $FeSCN^{2+}$ and H^+, what happens to the conc. of Fe^{3+}? How are the numbers of moles $FeSCN^{2+}$ produced and the number of moles Fe^{3+} used up related to each other?

2.   In an experiment similar to the one you will be doing, the $FeSCN^{2+}$ *equilibrium* concentration was found to be $0.32 \times 10^{-4} M$ in a solution made by mixing 10 cm^3 of 0.001 M $Fe(NO_3)_3$ with 10 cm^3 $0.001 M$ HSCN. The H^+ concentration was maintained at $0.5 M$.

   a. How many moles $FeSCN^{2+}$ are present at equilibrium?   _____ $\times 10^{-6}$

   b. How many moles of Fe^{3+} and HSCN were initially present?   _____ $\times 10^{-6}$

   c. How many moles Fe^{3+} remain unreacted in the solution?   _____ $\times 10^{-6}$

   d. How many moles HSCN remain unreacted?   _____ $\times 10^{-6}$

   e. Find $[Fe^{3+}]$ and [HSCN] in the equilibrium solution.   _____ $\times 10^{-4}M$

   f. Calculate the value of K_c for the reaction.

                                                      _____

3.   In this experiment we assume that the complex ion formed is $FeSCN^{2+}$. It would be possible, however, to form $Fe(SCN)_2^+$ under certain conditions.

   a. Write the equation for the reaction between Fe^{3+} and HSCN in which $Fe(SCN)_2^+$ is produced.

   b. Formulate the expression for the equilibrium constant K_c associated with the reaction in a.

CHEMICAL KINETICS

EXPERIMENT

20 • Rates of Chemical Reactions, I. The Bromination of Acetone

The rate at which a chemical reaction occurs depends on several factors: the nature of the reaction, the concentrations of the reactants, the temperature, and the presence of possible catalysts. All of these factors can markedly influence the observed rate of reaction.

Some reactions at a given temperature are very slow indeed; the oxidation of gaseous hydrogen or wood at room temperature would not appreciably proceed in a century. Other reactions are essentially instantaneous; the precipitation of silver chloride when solutions containing silver ions and chloride ions are mixed and the formation of water when acidic and basic solutions are mixed are examples of extremely rapid reactions. In this experiment we will study a reaction which, in the vicinity of room temperature, proceeds at a moderate, relatively easily measured rate.

For a given reaction, the rate typically increases with an increase in the concentration of any reactant. The relation between rate and concentration is a remarkably simple one in many cases, and for the reaction

$$aA + bB \rightarrow cC$$

the rate can usually be expressed by the equation

$$\text{rate} = k\,(A)^m\,(B)^n \tag{1}$$

where m and n are generally, but not always, integers, 0, 1, 2, or possibly 3; (A) and (B) are the concentrations of A and B (in moles per cubic decimetre); and k is a constant, called the *rate constant* of the reaction, which makes the relation quantitatively correct. The numbers m and n are called the *orders of the reaction* with respect to A and to B. If m is 1 the reaction is said to be *first order* with respect to the reactant A. If n is 2 the reaction is second order with respect to reactant B.

The rate of a reaction is also significantly dependent on the temperature at which the reaction occurs. An increase in temperature increases the rate, an often cited rule being that a 10° C rise in temperature will double the rate. This rule is only approximately correct; nevertheless, it is clear that a rise of temperature of say 100° C could change the rate of a reaction appreciably.

As with the concentration, there is a quantitative relation between reaction rate and temperature, but here the relation is somewhat more complicated. This relation is based on the idea that in order to react, the reactant species must have a certain minimum amount of energy present at the time the reactants collide in the reaction step; this amount of energy, which is typically furnished by the kinetic energy of motion of the species present, is called the *activation energy* for the reaction. The equation relating the rate constant k to the absolute temperature T and the activation energy E_a is

$$\log_{10} k = \frac{-E_a}{2.30RT} + \text{constant} \tag{2}$$

where R is the gas constant (8.31 J/mol · K for E_a in joules per mole). This equation is identical in form to Equation 1 in Exp. 17. By measuring k at different temperatures we can determine graphically the activation energy for a reaction.

In this experiment we will study the kinetics of the reaction between bromine and acetone:

$$\underset{\substack{\| \\ \text{CH}_3-\text{C}-\text{CH}_3(\text{aq})}}{\overset{\text{O}}{}} + \text{Br}_2(\text{aq}) \rightarrow \underset{\substack{\| \\ \text{CH}_3-\text{C}-\text{CH}_2\text{Br}(\text{aq})}}{\overset{\text{O}}{}} + \text{H}^+(\text{aq}) + \text{Br}^-(\text{aq})$$

The rate of this reaction is found to depend on the concentration of hydrogen ion in the solution as well as presumably on the concentrations of the two reactants. By Equation 1, the rate law for this reaction is

$$\text{rate} = k(\text{acetone})^m (\text{Br}_2)^n (\text{H}^+)^p \tag{3}$$

where m, n, and p are the orders of the reaction with respect to acetone, bromine, and hydrogen ion respectively, and k is the rate constant for the reaction.

The rate of this reaction can be expressed as the (small) change in the concentration of Br_2, $\Delta (\text{Br}_2)$, which occurs, divided by the time interval Δt required for the change:

$$\text{rate} = \frac{\Delta (\text{Br}_2)}{\Delta t} \tag{4}$$

Ordinarily, since rate varies as the concentrations of the reactants according to Equation 3, in a rate study it would be necessary to measure, directly or indirectly, the concentration of each reactant as a function of time; the rate would typically vary markedly with time, decreasing to very low values as the concentration of at least one reactant becomes very low. This makes reaction rate studies relatively difficult to carry out and introduces mathematical complexities that are difficult for beginning students to understand.

The bromination of acetone is a rather atypical reaction, in that it can be very easily investigated experimentally. First of all, bromine has color, so that one can readily follow changes in bromine concentration visually. A second and very important characteristic of this reaction is that it turns out to be zero order in Br_2 concentration. This means (see Equation 3) that the rate of the reaction does not depend on (Br_2) at all; $(\text{Br}_2)^0 = 1$, no matter what the value of (Br_2) is, as long as it is not itself zero.

Since the rate of the reaction does not depend on (Br_2), we can study the rate by simply making Br_2 the limiting reagent present in a large excess of acetone and H^+ ion. We then measure the time required for a known initial concentration of Br_2 to be completely used up. If both acetone and H^+ are present at much higher concentrations than that of Br_2, their concentrations will not change appreciably during the course of the reaction, and the rate will remain, by Equation 3, effectively constant until all the bromine is gone, at which time the reaction will stop. Under such circumstances, if it takes t seconds for the color of a solution having an initial concentration of Br_2 equal to $(Br_2)_0$ to disappear, the rate of the reaction, by Equation 4, would be

$$\text{rate} = \frac{\Delta\,(Br_2)}{\Delta\,t} = \frac{(Br_2)_0}{t} \tag{5}$$

Although the rate of the reaction is constant during its course under the conditions we have set up, we can vary it by changing the initial concentrations of acetone and H^+ ion. If, for example, we should *double* the initial concentration of *acetone* over that in Reaction 1, keeping (H^+) and (Br_2) at the *same* values they had previously, then the rate of Reaction 2 would, according to Equation 3, be different from that in Reaction 1:

$$\text{rate } 2 = k(2A)^m\,(Br_2)^0\,(H^+)^p$$
$$\text{rate } 1 = k(A)^m\,(Br_2)^0\,(H^+)^p$$

Dividing the first equation by the second, we see that the k's cancel, as do the terms in the bromine and hydrogen ion concentrations, since they have the same values in both reactions, and we obtain simply

$$\frac{\text{rate } 2}{\text{rate } 1} = \frac{(2A)^m}{(A)^m} = \left(\frac{2A}{A}\right)^m = 2^m \tag{6}$$

Having measured both rate 2 and rate 1 by Equation 5, we can find their ratio, which must be equal to 2^m. We can then solve for m either by inspection or using logarithms and so find the *order* of the reaction with respect to acetone.

By a similar procedure we can measure the order of the reaction with respect to H^+ ion concentration and also confirm the fact that the reaction is zero order with respect to Br_2. Having found the order with respect to each reactant, we can then evaluate k, the rate constant for the reaction.

The determination of the orders m and p, the confirmation of the fact that n, the order with respect to Br_2, equals zero, and the evaluation of the rate constant k for the reaction at room temperature comprise your assignment in this experiment. You will be furnished with standard solutions of acetone, bromine, and hydrogen ion, and with the composition of one solution that will give a reasonable rate. The rest of the planning and the execution of the experiment will be your responsibility.

An optional part of the experiment is to study the rate of this reaction at different temperatures in order to find its activation energy. The general procedure here would be to study the rate of reaction in one of the mixtures at room temperature and at two other temperatures, one above and one below room temperature. Knowing the rates, and hence the k's, at the three temperatures, you can then find E_a, the energy of activation for the reaction, by plotting $\log k$ vs. $1/T$. The slope of the resultant straight line, by Equation 2, must be $-E_a/2.30R$.

EXPERIMENTAL PROCEDURE

WEAR YOUR SAFETY GLASSES WHILE PERFORMING THIS EXPERIMENT

Select two 15 cm test tubes; when filled with distilled water, they should appear to have identical color when you view them down the tubes against a white background.

Draw 50 cm³ of each of the following solutions into clean dry 100 cm³ beakers, one solution to a beaker: 4 M acetone, 1 M HCl, and 0.02 M Br_2. Cover each beaker with a watch glass.

With your graduated cylinder, measure out 10.0 cm³ of the 4 M acetone solution and pour it into a clean 125 cm³ Erlenmeyer flask. Then measure out 10.0 cm³ 1 M HCl and add that to the acetone in the flask. Add 20.0 cm³ distilled H_2O to the flask. Drain the graduated cylinder, shaking out any excess water, and then use the cylinder to measure out 10.0 cm³ 0.02 M Br₂ solution. Be careful not to spill the bromine solution on your hands or clothes.

Noting the time on your wrist watch or the wall clock to one second, pour the bromine solution into the Erlenmeyer flask and quickly swirl the flask to thoroughly mix the reagents. The reaction mixture will appear yellow because of the presence of the bromine, and the color will fade slowly as the bromine reacts with the acetone. Fill one of the test tubes ¾ full with the reaction mixture, and fill the other test tube to the same depth with distilled water. Stopper the Erlenmeyer flask. Look down the test tubes toward a well-lit piece of white paper, and note the time when the color of the bromine just disappears. Measure the temperature of the mixture in the test tube.

Repeat the experiment, using as a reference the reacted solution instead of distilled water. The amount of time required in the two runs should agree within about thirty seconds.

The rate of the reaction equals the initial concentration of Br₂ *in the reaction mixture* divided by the elapsed time. Since the reaction is zero order in Br₂, and since both acetone and H⁺ ion are present in great excess, the rate is constant throughout the reaction and the concentrations of both acetone and H⁺ remain essentially at their initial values.

Having found the rate of the reaction for one composition of the system, change the composition of the reaction mixture by changing the volume, and hence the concentration, of acetone, keeping the total volume at 50 cm³, so that by measuring the rate of reaction in the new mixture you can find the order of the reaction with respect to acetone. (Remember that the concentrations of H⁺ and Br₂ must be the same as in the initial experiment!) Repeat the experiment with this mixture; the times required for reaction should not differ by more than about twenty seconds. Keep the temperature within about a degree of that in the first experiment. Calculate the rate of the reaction. Compare the rate with that found in the first mixture, and then calculate m, the order of the reaction with respect to acetone, using an equation similar to (6).

Again change the composition of the reaction mixture so that this time a measurement of the reaction rate will give you information about the order of the reaction with respect to H⁺. Repeat the experiment with this mixture to establish the time of reaction to within twenty seconds, again making sure that the temperature is within about a degree of that observed previously. From the rate you determine for this mixture find p, the order of the reaction with respect to H⁺.

Finally, change the reaction mixture composition in such a way as to allow you to show that the order of the reaction with respect to Br₂ is zero. Measure the rate of the reaction twice, and calculate n, the order with respect to Br₂.

Having found the order of the reaction for each species on which the rate depends, evaluate k, the rate constant for the reaction, from the rate and concentration data in each of the mixtures you studied. If the temperatures at which the reactions were run are all equal to within a degree or two, k should be about the same for each mixture.

As a final reaction, make up a mixture using reactant volumes that you did not use in any previous experiments. Using Equation 3, the values of concentrations in the mixtures, the orders, and the rate constant you calculated from your experimental data, predict how long it will take for the Br₂ color to disappear from your mixture. Measure the time for the reaction and compare it with your prediction.

If time permits, select one of the reaction mixtures you have already used which gave a convenient time, and use that mixture to measure the rate of reaction at about 10° C and at about 40° C. From the two rates you find, plus the rate at room temperature, calculate the energy of activation for the reaction, using Equation 2.

CAUTION: The reagents used in this experiment are volatile. Avoid exposure to their vapors by keeping the reagents and reaction mixtures covered, and, if convenient, opening the windows in the lab.

DATA AND CALCULATIONS: The Bromination of Acetone

I. Reaction Rate Data

Mixture	Volume in cm³ 4 M acetone	Volume in cm³ 1 M HCl	Volume in cm³ 0.02 M Br_2	Volume in cm³ H_2O	Time for Reaction in seconds 1st Run	2nd Run	Temp °C
I	10	10	10	20	_____	_____	_____
II	_____	_____	_____	_____	_____	_____	_____
III	_____	_____	_____	_____	_____	_____	_____
IV	_____	_____	_____	_____	_____	_____	_____
V	_____	_____	_____	_____	_____	_____	_____

II. Determination of Reaction Orders with Respect to Acetone, H^+ Ion, and Br_2

$$\text{rate} = k \, (\text{acetone})^m \, (Br_2)^n \, (H^+)^p$$

Mixture	(acetone)	(H^+)	$(Br_2)_0$	rate $= \dfrac{(Br_2)_0}{\text{ave. time}}$
I	0.8 M	0.2 M	0.004 M	_____
II	_____	_____	_____	_____
III	_____	_____	_____	_____
IV	_____	_____	_____	_____

$$\frac{\text{rate II}}{\text{rate I}} = \underline{\hspace{2cm}} = \left(\underline{\hspace{2cm}}\right)^m; \ m = \underline{\hspace{2cm}}$$

$$\frac{\text{rate III}}{\text{rate}\underline{\ \ }} = \underline{\hspace{2cm}} = \left(\underline{\hspace{2cm}}\right)^p; \ p = \underline{\hspace{2cm}}$$

Continued on following page **155**

$$\frac{\text{rate IV}}{\text{rate}\underline{\quad}} = \underline{\hspace{4cm}} = \left(\underline{\hspace{3cm}}\right)^{n}; \ n = \underline{\hspace{3cm}}$$

III.   Determination of Rate Constant k

Mixture   I   II   III   IV

k _____   _____   _____   _____   _____   average

IV.   Prediction of Reaction Rate

Reaction mixture

Volume 4 M acetone	Volume 1 M HCl	Volume 0.02 M Br_2	Volume H_2O	
_____	_____	_____	_____	_____

(acetone) _____ M (H^+) _____ M (Br_2) _____ M

Predicted rate _____

Predicted time for reaction _____ s

Observed time for reaction _____ s

V.   Determination of Energy of Activation (Optional)

Reaction mixture used _____ (same for all temperatures)

Time for reaction at about 10°C   _____ s   temperature _____ °C

Time for reaction at about 40°C   _____ s   temperature _____ °C

Time for reaction at room temp.   _____ s   temperature _____ °C

Continued on following page

Calculate the rate constant at each temperature from your data, following the procedure in III.

	rate	k	log k	$\dfrac{1}{T(K)}$
~10°C	_____	_____	_____	_____
~40°C	_____	_____	_____	_____
room temp.	_____	_____	_____	_____

Plot log k vs. $1/T$. Find the slope of the best straight line through the points.

Slope = _____

By Equation 2:

$$E_a = -(2.30)(8.31)(\text{slope})$$

$E_a =$ _____ J

THE BROMINATION OF ACETONE
(Data and Calculations)

$\log K$

−3.0

−4.0

−5.0

3.0×10^{-3} 3.5×10^{-3} 4.0×1

$1/T$

ADVANCE STUDY ASSIGNMENT: The Bromination of Acetone

1. In a reaction involving the bromination of acetone, the following initial concentrations were present in the reaction mixture:

$$(\text{acetone}) = 0.8 \ M; \quad (H^+) = 0.2 \ M; \quad (Br_2) = 0.004 \ M$$

In the reaction at 25°C, it took 240 s before the Br_2 had disappeared from the solution. If the reaction is zero order in Br_2, how long would it take before the color of Br_2 is discharged from a reaction mixture at 25°C exactly like the one above except that the concentration was initially 0.008 M in Br_2?

_____ s

2. The following data were obtained at 27°C for the reaction:

$$aA + bB \longrightarrow cC$$

Reaction mixture	1	2	3	4
Initial (A)	0.200	0.400	0.200	0.100
Initial (B)	0.200	0.200	0.400	0.300
Initial rate	0.75	3.00	1.50	0.28

What is the order of the reaction with respect to A? _____

What is the order of the reaction with respect to B? _____

3.   a. What volume of each of the solutions listed below would you use to prepare 50 cm^3 of the reaction mixture referred to in Problem 1?

_____ cm^3 4 M acetone   _____ cm^3 1 M HCl

_____ cm^3 0.020 M Br_2 _____ cm^3 water

   b. How would you make up 50 cm^3 of a mixture in which the concentrations of acetone, HCl, and Br_2 are 1.6 M, 0.2 M, and 0.004 M respectively?

_____ cm^3 4 M acetone   _____ cm^3 1 M HCl

_____ cm^3 0.020 M Br_2 _____ cm^3 water

EXPERIMENT

21 • Rates of Chemical Reactions, II. A Clock Reaction

In the previous experiment we discussed the factors that influence the rate of a chemical reaction and presented the terminology used in quantitative relations in studies of the kinetics of chemical reactions. That material is also pertinent to this experiment and should be studied before you proceed further.

This experiment involves the study of the rate properties, or chemical kinetics, of the following reaction between iodide ion and peroxydisulfate ion:[*]

$$2\,I^-(aq) + S_2O_8^{2-}(aq) \rightarrow I_2(aq) + 2\,SO_4^{2-}(aq) \tag{1}$$

This reaction proceeds reasonably slowly at room temperature, its rate depending on the concentrations of the I^- and $S_2O_8^{2-}$ ions according to the rate law discussed in the previous experiment. For this reaction the rate law takes the form:

$$\text{rate} = k(I^-)^m\,(S_2O_8^{2-})^n \tag{2}$$

One of the main purposes of the experiment will be to evaluate the rate constant k and the reaction orders m and n for this reaction. We will also investigate the manner in which the reaction rate depends on temperature and will evaluate the activation energy E_a for the reaction. Finally we shall briefly examine the effect of a catalyst on the rate of the reaction.

Our method for measuring the rate of the reaction involves what is frequently called a "clock" reaction. In addition to Reaction 1, whose kinetics we will study, the following reaction will also be made to occur simultaneously in the reaction flask:

$$I_2(aq) + 2\,S_2O_3^{2-}(aq) \rightarrow 2\,I^-(aq) + S_4O_6^{2-}(aq) \tag{3}$$

As compared with (1) this reaction is essentially instantaneous. The I_2 produced in (1) reacts completely with the thiosulfate, $S_2O_3^{2-}$, ion present in the solution, so that until all the thiosulfate ion has reacted, the concentration of I_2 is effectively zero. As soon as the $S_2O_3^{2-}$ is gone from the system, the I_2 produced by (1) remains in the solution and its concentration begins to increase. The presence of I_2 is made strikingly apparent by a starch indicator which is added to the reaction mixture, since I_2 even in small concentrations reacts with starch solution to produce a deep blue color.

By carrying out Reaction 1 in the presence of $S_2O_3^{2-}$ ion and a starch indicator, we introduce a built-in "clock" into the system. Our clock tells us when sufficient I_2 has been produced by Reaction 1 to use up all the $S_2O_3^{2-}$ ion originally added. Since, however, one mole of I_2 is produced for each mole of $S_2O_8^{2-}$ reacted, and each mole of I_2 reacts in (3) with two moles of $S_2O_3^{2-}$ ion, the color change also occurs at the time when a certain amount of $S_2O_8^{2-}$ ion has reacted, namely an amount in moles equal to one-half the number of moles $S_2O_3^{2-}$ initially present in the reaction flask. If we fix the amount of $S_2O_3^{2-}$ ion used at a value that is small compared to the amount of I^- and $S_2O_8^{2-}$ used, the color change will occur before any appreciable amounts of reactants are used up, and the concentrations of reactants and the rate in (2) will remain essentially constant during the time interval over which the rate is measured.

---

[*]Similar to Experiment 25 in Chemical Principles in the Laboratory, by H. W. Franz and L. E. Malm, Freeman, 1966.

The reaction between I^- and $S_2O_8^{2-}$ ions will be conducted under the conditions in the discussion above. Carefully measured amounts of I^- and $S_2O_8^{2-}$ ion in water solution will be mixed in the presence of a relatively small amount of $S_2O_3^{2-}$ ion and a starch indicator. The time it takes for the solution to turn blue will be measured for several different solutions in which the amounts and hence concentrations of I^- and $S_2O_8^{2-}$ ions are varied, but in which the amount of $S_2O_3^{2-}$ ion is held constant. Essentially, what we will measure in each case is the time required for the concentration of the $S_2O_8^{2-}$ ion to decrease by a constant predetermined amount. Since the rate of a reaction is equal to minus the change (small) in concentration of a reactant divided by the time required for the change to occur, the experimental data will furnish the information needed to find the rate of reaction in each solution. The calculation of the value of the rate constant k and the orders of the reaction with respect to I^- and $S_2O_8^{2-}$ follow from the dependence of reaction rate on reactant concentrations. The calculation of the activation energy E_a for the reaction is made from data obtained on the dependence of the reaction rate on temperature.

EXPERIMENTAL PROCEDURE   WEAR YOUR SAFETY GLASSES WHILE PERFORMING THIS EXPERIMENT

A. Dependence of Reaction Rate on Concentration

TABLE OF REACTION MIXTURES AT ROOM TEMPERATURE

Reaction	Reaction Flask	50 ml Flask
1	20.0 cm³ 0.200 M KI	20.0 cm³ 0.100 M $(NH_4)_2S_2O_8$
2	10.0 cm³ 0.200 M KI 10.0 cm³ 0.200 M KCl	20.0 cm³ 0.100 M $(NH_4)_2S_2O_8$
3	20.0 cm³ 0.200 M KI	10.0 cm³ 0.100 M $(NH_4)_2S_2O_8$ 10.0 cm³ 0.100 M $(NH_4)_2SO_4$
4	20.0 cm³ 0.200 M KI	5.0 cm³ 0.100 M $(NH_4)_2S_2O_8$ 15.0 cm³ 0.100 M $(NH_4)_2SO_4$
5	8.0 cm³ 0.200 M KI 12.0 cm³ 0.200 M KCl	20.0 cm³ 0.100 M $(NH_4)_2S_2O_8$
6	15.0 cm³ 0.200 M KI 5.0 cm³ 0.200 M KCl	15.0 cm³ 0.100 M $(NH_4)_2S_2O_8$ 5.0 cm³ 0.100 M $(NH_4)_2SO_4$

This table summarizes the volumes of reactants to be used in making up six different reaction mixtures. The actual procedure for carrying out each reaction will be much the same, and we will describe it now for Reaction 1.

Pipet 20.0 cm³ of 0.200 M KI into a 250 cm³ Erlenmeyer flask, which we will call the reaction flask and which will serve as the container for the reaction. Pipet 10.0 cm³ of 0.005 M $Na_2S_2O_3$ into this flask and add 3 or 4 drops of starch solution. Pipet 20.0 cm³ of 0.100 M $(NH_4)_2S_2O_8$ into a small 50 cm³ Erlenmeyer flask.

Put a thermometer into the reaction flask. Have a watch available with a second hand or use the second hand on the wall clock.

Pour the solution from the 50 cm³ flask into the reaction flask and swirl the solutions to mix them thoroughly. Note the time at which the solutions were mixed. Continue swirling the solution; it should turn blue in less than a minute. Record the time at the instant the blue color appears. Also record the temperature of the solution to ±0.2° C.

Repeat the experiment with the other mixtures in the table. In each case use 10.0 ml of 0.005 M $Na_2S_2O_3$ solution and a few drops of starch indicator in the reaction flask. The flasks used should be rinsed with distilled water between experiments and drained before being used again. Pipets should be used in measuring the volumes of KI, $(NH_4)_2S_2O_8$, and $Na_2S_2O_3$ solutions. The volumes of the KCl and $(NH_4)_2SO_4$ solutions can be measured out with a graduated cylinder. These two solutions are used rather than water in diluting the reaction mixture so that the ionic strength of the mixture, which has some effect on reaction rate, can be kept substantially constant. The temperature should be kept the same for all experiments. Repeat any experiments that for any reason did not appear to proceed properly.

B. Dependence of Reaction Rate on Temperature.   In this part of the experiment the reaction will be carried out at several different temperatures, but with the same concentrations of all reactants. The Table indicates the reaction conditions.

TABLE OF REACTION MIXTURES AT DIFFERENT TEMPERATURES

Reaction	Temperature	Reaction Mixture
1	about 20°C	as in Reaction 1
7	about 40°C	as in Reaction 1
8	about 10°C	as in Reaction 1
9	about 0°C	as in Reaction 1

The rate of the reaction at about 20° C can be obtained from the data on Reaction 1.

Reaction 7 is carried out by first pipetting out the same solutions in the same volumes as those used in Reaction 1: 20.0 cm³ of 0.200 M KI, 10.0 cm³ of 0.005 M $Na_2S_2O_3$, and a few drops of starch indicator into the reaction flask, and 20.0 cm³ of 0.100 M $(NH_4)_2S_2O_8$ into the smaller flask.

Instead of mixing the solutions at room temperature, put the two flasks in water at 40° C in one or more large beakers. Make sure that the water in the large beakers is all at about 40° C, and leave the flasks in the water for several minutes so that they will be at the proper temperature. Put the thermometer in the reaction flask, and when the temperature is about 40° C, mix the solutions together, noting the time of mixing. Swirl as before, keeping the reacting mixture in the warm water. Record the time at which the color change occurs and the temperature of the mixture at that point.

Repeat the experiment at about 10° C, cooling all reactants in a water bath to that temperature before starting the reaction. Record the time required for the reaction and the final temperature of the reaction mixture. Repeat once again at about 0° C, this time using an ice-water bath to cool the reactant solutions.

C. Dependence of Reaction Rate on the Presence of a Catalyst (Optional).   Metallic cations have a pronounced catalytic effect on the rate of this reaction. Observe this effect by repeating Reaction 1 with a catalyst at room temperature. Before mixing the three solutions, add five drops of 0.1 M $CuSO_4$ to the flask containing the 0.1 M $(NH_4)_2S_2O_8$. Swirl the flask for a minute to mix the catalyst thoroughly. Then mix the solutions, and when the color change occurs record the time and temperature.

DATA AND CALCULATIONS: Rates of Chemical Reactions, II.
A Clock Reaction

A. Orders of the Reaction. Rate constant determination

$$\text{Reaction:} \quad 2\,I^-(aq) + S_2O_8^{2-}(aq) \rightarrow I_2(aq) + 2\,SO_4^{2-}(aq) \tag{1}$$

$$\text{rate} = k(I^-)^m\,(S_2O_8^{2-})^n = \frac{-\Delta(S_2O_8^{2-})}{t} \tag{4}$$

In all the reaction mixtures used in the experiment, the color change occurred when a constant predetermined number of moles of $S_2O_8^{2-}$ had been used up by the reaction. The color "clock" allows you to measure the *time required* for this *fixed number of moles of* $S_2O_8^{2-}$ *to react*. The rate of each reaction is determined by the time t required for the color to change; since in Equation 4 the change in concentration of $S_2O_8^{2-}$ ion, $\Delta(S_2O_8^{2-})$, is the same in each mixture, the relative rate of each reaction is inversely proportional to the time t. Since we are mainly concerned with relative rather than absolute rate, we will for convenience take all relative rates as being equal to $100/t$. Fill in the table below, calculating reactant concentrations and relative reaction rate for each mixture.

Reaction	Time t (s) for Color to Change	Initial Concentrations in Reaction Flask I^-	$S_2O_8^{2-}$	Relative Rate of Reaction, $100/t$
1	_____	_____	_____	_____
2	_____	_____	_____	_____
3	_____	_____	_____	_____
4	_____	_____	_____	_____
5	_____	_____	_____	_____
6	_____	_____	_____	_____

Temperature of Reaction 1 _____ °C

Consider the relative rates of Reactions 1 to 6. These rates differ because the concentrations of I^- and $S_2O_8^{2-}$ ions differ in the various reaction mixtures. We can relate the rates of each of these reactions to the concentrations by modifying Equation 2 to read

$$\text{Relative rate} = k'(I^-)^m\,(S_2O_8^{2-})^n \tag{5}$$

where k' is a relative rate constant. The problem now is to find values for m, n, and k' such that the data in the above table are consistent with Equation 5.

The solution to this problem is not as difficult as it might first appear. Note that the concentrations of I^- and $S_2O_8^{2-}$ ions change in a simple manner, as one goes from one mixture to the next; one concentration remains constant while the other changes by a factor of 2 or a rational fraction. This means that Equation 5 for one reaction mixture can be related to Equation 5 for another in such a way as will permit easy evaluation of m and n.

165

Continued on following page

Write Equation 5 below for Reactions 1 and 2, substituting the known concentrations of I^- and $S_2O_8^{2-}$ ions:

Relative rate 1 = _____ = $k'($ $)^m ($ $)^n$

Relative rate 2 = _____ = $k'($ $)^m ($ $)^n$

Divide the first equation by the second:

$$=$$

If you have done this properly, you will have an equation involving only m as an unknown. Solve this equation for m, the order of the reaction with respect to I^- ion.

$$m = \text{_____}\text{(ordinarily an integer)}$$

Applying the same approach to Reactions 1 and 3, find the value of n, the order of the reaction with respect to $S_2O_8^{2-}$ ion.

Relative rate 1 = _____ = $k'($ $)^m ($ $)^n$

Relative rate 3 = _____ = $k'($ $)^m ($ $)^n$

Dividing one equation by the other:

$$=$$

$$n = \text{_____}$$

Having found m and n, the relative rate constant, k', can be calculated by substitution of m, n, and the known rates and reactant concentrations into Equation 5. Evaluate k' for Reactions 1 to 5.

Reaction	1	2	3	4	5
k'	_____	_____	_____	_____	_____ k'_{ave} _____

Why should k' have nearly the same value for each of the above Reactions?

Using k'_{ave}, predict the relative rate and time t for Reaction 6.

relative rate$_{pred}$ _____ t_{pred} _____ t_{obs} _____

Comment on the validity of the reaction rate law (5) and hence (2) for the mixtures studied.

Continued on following page

B. The Effect of Temperature on Reaction Rate: The Activation Energy. To find the activation energy for the reaction you will need to complete the table below, using the procedure indicated if necessary.

The dependence of the rate constant, k', for a reaction is given by Equation 2 in Experiment 20:

$$\log_{10} k' = \frac{-E_a}{2.3\,RT} + \text{constant} \tag{6}$$

where the terms in the equation have the meanings given in the discussion in that experiment. To set up the terms in $1/T$, fill in (b), (e), and (f) in the table.

Since the reactions in mixtures 1, 7, 8, and 9 all involve the same reactant concentrations, the rate constants, k', for two different mixtures will have the same ratio as the reaction rates themselves for the two mixtures. This means that in the calculation of E_a, we can use the observed relative rates instead of rate constants. Proceeding as before, calculate the relative rates of reaction in each of the mixtures and enter these values in (c) below. Take the $\log_{10}$ rate for each mixture and enter these values in (d).

	Reaction			
	1	7	8	9
(a) Time t in seconds for color to appear	____	____	____	____
(b) Temperature of the reaction mixture in °C	____	____	____	____
(c) Relative rate = $100/t$	____	____	____	____
(d) $\log_{10}$ of relative rate	____	____	____	____
(e) Temperature T in K	____	____	____	____
(f) $1/T$, K^{-1}	____	____	____	____

To evaluate E_a, make a graph of log relative rate vs. $1/T$ on the graph paper provided. Find the slope of the line obtained by drawing the best straight line through the experimental points.

Slope = _____

The slope of the line equals $-E_a/2.3\,R$, where $R = 8.31$ J/mol · K if E_a is to be in joules per mole. Find the activation energy, E_a, for the reaction.

E_a = _____ J

Continued on following page

C. Effect of a Catalyst on Reaction Rate

	Reaction 1	Catalyzed Reaction 1
Time for color to appear (seconds)	_____	_____

Would you expect the activation energy, E_a, for the catalyzed reaction to be greater than, less than, or equal to the activation energy for the uncatalyzed reaction? Why?

RATES OF CHEMICAL REACTIONS, II. A CLOCK REACTION
(Data and Calculations)

log relative rate

0.00300   0.00350   0.00400

1/T

ADVANCE STUDY ASSIGNMENT: Rate of Chemical Reactions, II.
A Clock Reaction

1. In a clock reaction 20 cm³ of 0.200 M KI, 20 cm³ of 0.100 M $(NH_4)_2S_2O_8$, and 10 cm³ of 0.005 M $Na_2S_2O_3$ were mixed in the presence of a starch indicator. The solution turned blue in 88 s.

   a. What were the concentrations of I^-, $S_2O_8^{2-}$, and $S_2O_3^{2-}$ in the reaction mixture?

      conc. I^- = _____M; conc. $S_2O_8^{2-}$ = _____M;

      conc. $S_2O_3^{2-}$ = _____M

   b. What fraction of the $S_2O_8^{2-}$ had reacted when the mixture turned blue?

      _____

   c. What was the initial relative rate (see page 165)?

      _____

2. If $(2.0/1.0)^m = 2.2$, what is the approximate value of m? The exact value? (Hint: take logarithms of both sides of the equation.)

   _____ approx.; _____ exact

3. In the reaction A + B → C, it took 15 s for 1 per cent of B to react in a certain mixture of A and B. When the concentration of A in the mixture was decreased by a factor of 2, keeping the concentration of B constant, it took 32 s for 1 per cent of B to react at the same temperature. What is the order of the reaction with respect to A?

   _____ order with respect to A

PRECIPITATION REACTIONS

EXPERIMENT

22 • Determination of an Unknown Chloride

One of the important applications of precipitation reactions lies in the area of quantitative analysis. Many substances that can be precipitated from solution are so slightly soluble that the precipitation reaction by which they are formed can be considered to proceed to completion. Silver chloride is an example of such a substance.

$$AgCl(s) \rightleftharpoons Ag^+(aq) + Cl^-(aq) \qquad K_{sp} = [Ag^+][Cl^-] = 1.6 \times 10^{-10}$$

Although silver chloride is in chemical equilibrium with its ions in solution, the equilibrium constant K_{sp} for the reaction is so low that if AgCl is precipitated by the addition of a solution containing Ag^+ ion to one containing Cl^- ion, essentially all the Ag^+ added will precipitate as AgCl until essentially all the Cl^- is used up. When the amount of Ag^+ added to the solution is equal to the amount of Cl^- originally present, the precipitation of Cl^- ion will be, for all practical purposes, complete.

A convenient method for chloride analysis using AgCl has been devised. A solution of $AgNO_3$ is added to a chloride solution just to the point where the number of moles of Ag^+ added is equal to the number of moles of Cl^- initially present. We analyze for Cl^- by simply measuring how many moles of $AgNO_3$ are required. Surprisingly enough, this measurement is rather easily made by an experimental procedure called a *titration*.

In the titration a solution of $AgNO_3$ of known concentration (in moles $AgNO_3$ per cubic decimetre) is added from a calibrated buret to a solution containing a measured amount of unknown. The titration is stopped when a color change occurs in the solution, indicating that stoichiometrically equivalent amounts of Ag^+ and Cl^- are present. The color change is caused by a chemical reagent, called an indicator, which is added to the solution at the beginning of the titration.

The volume of $AgNO_3$ solution that has been added up to the time of the color change can be measured accurately with the buret, and the number of moles of Ag^+ added can be calculated from the known concentration of the solution.

In the Mohr method for the volumetric analysis of chloride, which we will employ in this experiment, the indicator used is K_2CrO_4. The chromate ion present in solutions of this substance will react with silver ion to form a red precipitate of Ag_2CrO_4. This precipitate will form as soon as $[Ag^+]^2 \times [CrO_4^{2-}]$ exceeds the solubility product of Ag_2CrO_4, which is about 1×10^{-12}. Under the conditions of the titration, the Ag^+ added to the solution reacts preferentially with Cl^- until that ion is essentially quantitatively removed

173

from the system, at which point Ag_2CrO_4 begins to precipitate and the solution color changes from yellow to buff. The end point of the titration is that point at which the color change is first observed.

In this experiment, weighed samples containing an unknown percentage of chloride will be titrated with a standardized solution of $AgNO_3$, and the volumes of $AgNO_3$ solution required to reach the end point of each titration will be measured. Given the molarity of the $AgNO_3$

$$\text{no. of moles } Ag^+ = \text{no. of moles } AgNO_3 = M_{AgNO_3} \times V_{AgNO_3}$$

where the volume of $AgNO_3$ is expressed in cubic decimetres and the molarity M_{AgNO_3} is in moles per cubic decimetre. At the end point of the titration,

$$\text{no. of moles } Ag^+ \text{ added} = \text{no. of moles } Cl^- \text{ present in unknown}$$

$$\text{no. of grams } Cl^- \text{ present} = \text{no. of moles } Cl^- \text{ present} \times GAM \text{ Cl}$$

$$\% \text{ Cl} = \frac{\text{no. of grams } Cl^-}{\text{no. of grams unknown}} \times 100$$

EXPERIMENTAL PROCEDURE

Obtain from the stockroom a buret and a vial containing a sample of an unknown solid chloride. Weigh out accurately on the analytical balance three samples of the chloride, each sample weighing about 0.2 g. This weighing is best done by accurately weighing the vial and its contents and pouring out the sample a little at a time into a 250 cm³ Erlenmeyer flask until the vial has lost about 0.2 g of chloride sample. Again weigh the sample vial accurately to obtain the exact amount of chloride sample poured into the flask. Put two other samples of similar weight into clean, dry, small beakers, weighing the vial accurately after the size of each sample has been decided upon. Add 50 cm³ of distilled water to the flask to dissolve the sample and add 3 drops of some 1 M K_2CrO_4 indicator solution. Using the graduated cylinder at the reagent shelf, measure out about 100 cm³ of the standardized $AgNO_3$ solution into a clean *dry* 125 cm³ Erlenmeyer flask. This will be your total supply for the entire experiment so do not waste it. Clean your buret thoroughly with soap solution and rinse it with distilled water. Pour three successive 2 or 3 cm³ portions of the $AgNO_3$ solution into the buret and tip it back and forth to rinse the inside walls. Allow the $AgNO_3$ solution to drain out the buret tip completely each time. Fill the buret with the $AgNO_3$ solution. Open the buret stopcock momentarily to flush any air bubbles out of the tip of the buret. Be sure your stopcock fits snugly and that the buret does not leak.

Read the initial buret level to 0.02 cm³. You may find it useful when making readings to put a white card marked with a thick, black stripe behind the meniscus. If the black line is held just below the level to be read, its reflection in the surface of the meniscus will help you obtain a very accurate reading. Begin to add the $AgNO_3$ solution to the chloride solution in the Erlenmeyer flask. A white precipitate of AgCl will form immediately, and the amount will increase during the course of the titration. At the beginning of the titration, you can add the $AgNO_3$ fairly rapidly, a few ml at a time, swirling the flask as best you can to mix the solution. You will find that at the point where the $AgNO_3$ hits the solution, there will be a red spot of Ag_2CrO_4, which disappears when you stop adding nitrate and swirl the flask. As you proceed with the titration, the red spot will persist more and more, since the amount of excess chloride ion, which reacts with the Ag_2CrO_4 to form AgCl, will slowly decrease. Gradually decrease the rate at which you add $AgNO_3$ as the red color becomes stronger. At some stage you may find it convenient to set your buret stopcock to deliver $AgNO_3$ slowly, drop by drop, while you swirl the flask. When you are near the end point, add the $AgNO_3$ drop by drop, swirling between drops. The end point of the titration is that point where the mixture first takes on a permanent reddish-yellow or buff color which does not revert to pure yellow on swirling. If you are careful, you can hit the end point within one drop of $AgNO_3$. When you have reached the end point, stop the titration and record the buret level.

Rinse out the 250 cm³ Erlenmeyer flask in which you carried out the titration. Take your second sample and carefully pour it from the beaker into the Erlenmeyer flask. Wash out the beaker a few times with distilled water from your wash bottle and pour the washings into the sample flask. *All the sample* must be transferred if the analysis is to be accurate. Add water to the sample flask to a volume of about 50 cm³ and swirl to dissolve the solid. Refill your buret, take a volume reading, add the indicator, and proceed to titrate to an end point as before. This titration should be more accurate than the first, since the volume of $AgNO_3$ used is proportional to the sample size and can therefore be estimated rather well on the basis of the relative masses of the two samples.

Titrate the third sample as you did the second.

Name _____ Section _____

DATA: Determination of an Unknown Chloride

Unknown sample no. _____

Molarity of standard $AgNO_3$ solution _____

	I	II	III
Mass of vial and chloride unknown	_____ g	_____ g	_____ g
Mass of vial less sample	_____ g	_____ g	_____ g
Initial buret reading	_____ cm³	_____ cm³	_____ cm³
Final buret reading	_____ cm³	_____ cm³	_____ cm³

Calculations and Results

	I	II	III
Mass of sample	_____ g	_____ g	_____ g
Volume of $AgNO_3$ used to titrate sample	_____ cm³	_____ cm³	_____ cm³
No. of moles of $AgNO_3$ used to titrate sample	_____	_____	_____
No. of moles of Cl^- present in sample	_____	_____	_____
Mass of Cl^- present in sample	_____ g	_____ g	_____ g
% Cl^- in sample	_____ %	_____ %	_____ %
Mean value of % Cl^- in unknown	_____ %		

177

ADVANCE STUDY ASSIGNMENT: Determination of an Unknown Chloride

1.   A sample containing 0.175 g Cl^- is dissolved in 50 cm^3 of water. What is the concentration of Cl^- in moles per cubic decimetre?

_____M

2.   The solution in Problem 1 is titrated with $AgNO_3$ to an end point. If at that point $[Ag^+]$ equals $[Cl^-]$, as it should, find $[Cl^-]$ in the solution. If the final volume of the solution is 75 cm^3, how many grams of Cl^- are present in the solution? What per cent of the Cl^- originally present remains unprecipitated? $K_{sp}AgCl = 1.6 \times 10^{-10}$

_____M, _____ g, _____ %

3.   In the Mohr titration, the first formation of red Ag_2CrO_4 is taken to indicate the end point. In the solution being titrated, $[CrO_4^{2-}]$ is about 0.01 M. What is the $[Ag^+]$ when Ag_2CrO_4 first begins to precipitate? This concentration of Ag^+ is also in equilibrium with $AgCl(s)$, which is present in the system. What is $[Cl^-]$ in the solution when Ag_2CrO_4 first starts to appear? If the solution was initially 0.10 M in Cl^-, what fraction of Cl^- remains in solution at the Mohr end point? $K_{sp}Ag_2CrO_4 = 1 \times 10^{-12}$

$[Ag^+] =$ _____M, $[Cl^-] =$ _____M

_____ % in solution

4.   A solid chloride sample weighing 0.285 g required 47.28 cm^3 of 0.109 M $AgNO_3$ to reach the Ag_2CrO_4 end point. What is the per cent chloride in the sample?

_____%

23 • Determination of the Solubility Product of PbI$_2$

In this experiment you will determine the solubility product of lead iodide, PbI$_2$. Lead iodide is relatively insoluble, having a solubility of less than 0.002 mol/dm^3 at 20°C. The equation for the solution reaction of PbI$_2$ is

$$PbI_2(s) \rightleftarrows Pb^{2+}(aq) + 2\,I^-(aq) \tag{1}$$

The solubility product expression associated with this reaction is

$$K_{sp} = [Pb^{2+}]\,[I^-]^2 \tag{2}$$

Equation 2 implies that in any system containing solid PbI$_2$ in equilibrium with its ions, the product of $[Pb^{2+}]$ times $[I^-]^2$ will at a given temperature have a fixed magnitude, independent of how the equilibrium system was initially made up.

In the first part of the experiment, known volumes of standard solutions of Pb(NO$_3$)$_2$ and KI will be mixed in several different proportions. The yellow precipitate of PbI$_2$ formed will be allowed to come to equilibrium with the solution. The value of $[I^-]$ in the solution will be measured experimentally. The $[Pb^{2+}]$ will be calculated from the initial composition of the system, the measured value of $[I^-]$, and the stoichiometric relation between Pb^{2+} and I$^-$ in Equation 1.

In the second part of the experiment, a precipitate of PbI$_2$ will be prepared, washed clean of excess ions, and then dissolved in an inert salt solution. The $[I^-]$ in the saturated solution will be measured and the value of $[Pb^{2+}]$ found from the relation in Equation 1.

The concentration of I$^-$ ion will be found spectrophotometrically, as in Experiment 19. Although the iodide ion is not colored, it is relatively easily oxidized to I$_2$, which is brown in water solution. Our procedure will be to separate the solid PbI$_2$ from the solution and then to oxidize the I$^-$ in solution with potassium nitrite, KNO$_2$, under slightly acidic conditions, where the conversion to I$_2$ is quantitative. Although the concentration of I$_2$ will be rather low in the solutions you will prepare, the absorption of light by I$_2$ in the vicinity of 525 nm is sufficiently intense to make accurate analyses possible.

In all of the solutions prepared, potassium nitrate KNO$_3$ (note this distinction between KNO$_2$ and KNO$_3$!) will be present as an inert salt. This salt serves to keep the ionic strength of the solution essentially constant at 0.2 M and promotes the formation of well-defined crystalline precipitates of PbI$_2$.

EXPERIMENTAL PROCEDURE

WEAR YOUR SAFETY GLASSES WHILE PERFORMING THIS EXPERIMENT

Label five test tubes 1 to 5. Pipette 5.0 cm^3 of 0.012 M Pb(NO$_3$)$_2$ in 0.20 M KNO$_3$ into each of the first four test tubes. Add 2.0 cm^3 of 0.03 M KI in 0.20 M KNO$_3$ to test tube 1. Add 3.0, 4.0, and 5.0 cm^3 of this solution to test tubes 2, 3, and 4, respectively. Add enough 0.20 M KNO$_3$ to the first three test tubes to make the total volume 10.0 cm^3. Stopper each test tube and shake *thoroughly* at intervals of several minutes while you are proceeding with the next part of the experiment.

In a fifth test tube mix about 10 cm^3 of 0.012 M Pb(NO$_3$)$_2$ in KNO$_3$ with 10 cm^3 of 0.03 M KI in KNO$_3$. Shake the mixture vigorously for a minute or so. Let the solid settle for a few minutes and then decant and discard three-fourths of the supernatant solution.

Transfer the solid PbI$_2$ and the rest of the solution to a small test tube and centrifuge. Discard the liquid, retaining the solid precipitate. Add 3 cm^3 0.20 M KNO$_3$ and shake to wash the solid free of excess Pb^{2+} or I$^-$. Centrifuge again, and discard the liquid. By this procedure you should now have prepared a small sample of essentially pure PbI$_2$ in a little KNO$_3$ solution. Add 0.20 M KNO$_3$ to the solid until the tube is about three-fourths full. Shake well at several one minute intervals to saturate the solution with PbI$_2$.

In this experiment it is essential that the volumes of reagents used to make up the mixtures in test tubes 1 to 4 be measured accurately. It is also essential that all five mixtures be shaken thoroughly so that equilibrium can be established. Insufficient shaking of the first four test tubes will result in not enough PbI$_2$ precipitating to reach true equilibrium; if the small test tube is not shaken sufficiently, not enough PbI$_2$ will dissolve to attain equilibrium.

When each of the mixtures has been shaken for at least 15 min, let the tubes stand for three to four minutes to let the solid settle. Pour the supernatant liquid in test tube 1 into a small dry test tube and centrifuge for about three minutes to settle the solid PbI$_2$. Pour the liquid into another small dry test tube; if there are any solid particles or yellow color remaining in the liquid, centrifuge again. When you have a clear liquid, dip a small piece of clean, dry paper towel into the liquid to remove floating PbI$_2$ particles from the surface. Pipette 3.0 cm^3 of 0.20 M KNO$_2$ into a clean, dry spectrophotometer tube and add 2 drops 6 M HCl. Then add enough of the clear centrifuged solution (about 3 cm^3) to fill the spectrophotometer tube just to the level indicated by your instructor. Shake gently to mix the reagents and then measure the absorbance of the solution as directed by your instructor. The calibration curve which is provided will allow you to determine directly the concentration of I$^-$ ion that was in equilibrium with PbI$_2$. Use the same procedure to analyze the solutions in test tubes 2 through 5, completing each analysis before you proceed to the next.

DATA AND CALCULATIONS: Determination of the Solubility Product of PbI_2

From the experimental data we obtain $[I^-]$ directly. To obtain K_{sp} for PbI_2 we must calculate $[Pb^{2+}]$ in each equilibrium system. This is most easily done by constructing an equilibrium table. We first find the initial amounts of I^- and Pb^{2+} ions in each system from the way the mixtures were made up. Knowing $[I^-]$ and the formula of lead iodide allows us to calculate $[Pb^{2+}]$. K_{sp} then follows directly. The calculations are similar to those in Experiment 19.

$$PbI_2(s) \rightleftharpoons Pb^{2+}(aq) + 2\,I^-(aq) \qquad K_{sp} = [Pb^{2+}]\,[I^-]^2$$

DATA

Test tube no.	1	2	3	4	Saturated solution of PbI_2
cm^3 0.012 M $Pb(NO_3)_2$	_____	_____	_____	_____	
cm^3 0.03 M KI	_____	_____	_____	_____	
cm^3 0.20 M KNO_3	_____	_____	_____	_____	
Total volume (cm^3)	_____	_____	_____	_____	
Absorbance of solution	_____	_____	_____	_____	_____
$[I^-]$ (mol/dm³)	_____	_____	_____	_____	_____

CALCULATIONS

	1	2	3	4	
Initial no. moles Pb^{2+}	____$\times 10^{-5}$	____$\times 10^{-5}$	____$\times 10^{-5}$	____$\times 10^{-5}$	
Initial no. moles I^-	____$\times 10^{-5}$	____$\times 10^{-5}$	____$\times 10^{-5}$	____$\times 10^{-5}$	
No. moles I^- at equilibrium	____$\times 10^{-5}$	____$\times 10^{-5}$	____$\times 10^{-5}$	____$\times 10^{-5}$	
No. moles I^- precipitated	____$\times 10^{-5}$	____$\times 10^{-5}$	____$\times 10^{-5}$	____$\times 10^{-5}$	
No. moles Pb^{2+} precipitated	____$\times 10^{-5}$	____$\times 10^{-5}$	____$\times 10^{-5}$	____$\times 10^{-5}$	
No. moles Pb^{2+} at equilibrium	____$\times 10^{-5}$	____$\times 10^{-5}$	____$\times 10^{-5}$	____$\times 10^{-5}$	
$[Pb^{2+}]$ at equilibrium	_____	_____	_____	_____	_____
K_{sp} PbI_2	_____	_____	_____	_____	_____

ADVANCE STUDY ASSIGNMENT: Determination of the Solubility Product of PbI_2

1.   State in words the meaning of the solubility product equation for PbI_2:

$$K_{sp} = [Pb^{2+}][I^-]^2$$

2.   When 5.0 cm³ of 0.012 M $Pb(NO_3)_2$ are mixed with 5.0 cm³ of 0.030 M KI, a yellow precipitate of $PbI_2(s)$ forms.

   a. How many moles of Pb^{2+} are originally present?

   _____mol

   b. How many moles of I^- are originally present?

   _____mol

   c. In a colorimeter the equilibrium solution is analyzed for I^-, and its concentration is found to be 8×10^{-3} mol/dm³. How many moles of I^- are present in the solution (10 cm³)?

   _____mol

   d. How many moles of I^- precipitated?

   _____mol

   e. How many moles of Pb^{2+} precipitated?

   _____mol

   f. How many moles of Pb^{2+} are left in solution?

   _____mol

   g. What is the concentration of Pb^{2+} in the equilibrium solution?

   _____ M

   h. Find a value for K_{sp} of PbI_2 from these data.

   _____   **185**

24 • Preparation of Tin(II) Iodide

Certain combinations of cations and anions in aqueous solution will react and yield precipitates. Precipitation reactions are the basis of the preparation methods for many salts. In this experiment we will prepare a slightly soluble salt, tin(II) iodide, SnI_2, by mixing together a solution of a soluble tin(II) salt and a solution of a soluble iodide.

$$Sn^{2+}(aq) + 2\,I^-(aq) \rightarrow SnI_2(s)$$

A solution of Sn^{2+} can be readily obtained by dissolving a sample of tin foil of known weight in hydrochloric acid.

$$Sn(s) + 2\,H^+(aq) \rightarrow Sn^{2+}(aq) + H_2(g)$$

The iodide solution will be made by reducing iodine in water solution with zinc.

$$Zn(s) + I_2 \rightarrow Zn^{2+}(aq) + 2\,I^-(aq)$$

The precipitate of SnI_2 first obtained upon mixing the two solutions will be purified by recrystallization from solution in hydrochloric acid.

Your grade in this experiment will be based partly on the amount of SnI_2 you turn in and partly on its purity. Follow instructions carefully. You will be penalized if for any reason you need to get another sample of tin.

EXPERIMENTAL PROCEDURE

WEAR YOUR SAFETY GLASSES WHILE
PERFORMING THIS EXPERIMENT

Obtain a 4.5 g sample of tin foil, a Buchner funnel, and a filter flask from the storeroom. Tear the foil into small strips 2 to 4 cm wide. Place the tin strips in a 125 cm³ Erlenmeyer flask and add 20 cm³ of 12 M HCl (*CAUTION: Corrosive*) and 1 cm³ of 0.1 M $CuSO_4$ solution. Heat the flask under the hood with your Bunsen burner, keeping the solution simmering just below the boiling point so that the HCl does not boil away.

Place about 8 g of granulated zinc together with 10 cm³ water in a 250 cm³ Erlenmeyer flask. Weigh out in a 100 cm³ beaker on the triple-beam balance the amount of iodine that you have calculated in the Advance Study Assignment, and add it to the flask containing the zinc. Cool the flask with tap water to prevent the iodine from vaporizing as the reaction proceeds. When the reaction is complete the brown color of dissolved iodine will disappear, leaving a colorless or pale yellow solution. Decant the solution of ZnI_2 into a 250 cm³ beaker, being careful not to allow any excess zinc metal to get into the beaker. Rinse the Erlenmeyer flask with no more than 5 cm³ of water to get all of the ZnI_2 into the beaker. Put the unreacted zinc into the waste container.

Check the tin-HCl reaction mixture in the 125 cm³ flask. If considerable tin remains undissolved and hydrogen is still being generated, allow the reaction to continue. If there is tin left and no hydrogen is being evolved, add 5 cm³ more of 12 M HCl and continue to heat. A small residue of black particles that might remain at the bottom of the flask indicates the presence of some acid insoluble impurities in the tin.

When all of the tin has dissolved, decant the $SnCl_2$ solution into the beaker containing the zinc iodide solution. Rinse the black residue in the flask with 5 cm³ of water and add the rinsings to the beaker. Be careful not to transfer any solid material to the

ZnI_2 solution in this operation. A yellow precipitate of SnI_2, which quickly turns orange, should form immediately. Stir vigorously until you have a viscous mixture of solid with very little liquid.

The crude SnI_2 contained in the beaker must be separated from the impurities, mostly $ZnCl_2$, in the liquid mixed with the precipitate. First, the material in the beaker must be transferred to the Buchner funnel and filtered. Using your spatula, transfer as much of the slurry as you can to the funnel before turning on the suction.

Dry the precipitate on the filter as much as possible with suction. Discard the filtrate. Remove the filter paper and place the solid filter cake in a 250 cm³ beaker. Scrape the inside of the funnel and the filter paper to get as much of the product into the beaker as possible.

Add 5 cm³ of 12 M HCl and 40 cm³ of water to the solid in the beaker. Mix thoroughly and heat to a gentle boil, stirring occasionally.

While the solution is being heated, clean a 100 cm³ beaker and set up a conical filter arranged to drain into the beaker. If necessary, add one or two 5 cm³ portions of water to the boiling solution of SnI_2 in HCl in order to dissolve all or nearly all of the orange solid. When the SnI_2 has dissolved, warm the conical funnel by brushing the burner flame back and forth over the outside of the funnel, including the stem. The funnel should be a little too hot to touch, but *not* so hot that the paper ignites. Wrap the hot beaker of SnI_2 with a towel and pour the solution quickly through the conical filter. SnI_2 will begin to crystallize almost immediately from the solution draining into the beaker. Allow the filtrate to cool to room temperature and then place the beaker in an ice bath. Filter the cold contents of the beaker through the Buchner funnel to isolate the purified SnI_2. Use suction to dry the crystals as much as possible.

Turn off the suction and add 5 cm³ ice-cold distilled water to the funnel. Wait ten seconds and reapply suction for a minute or so. Transfer the solid to a dry piece of filter paper and press it dry with another sheet of filter paper. Let the SnI_2 dry in air a few minutes on a third filter paper and then weigh it on the paper to 0.1 g. Show your product to your instructor for his evaluation.

DATA AND RESULTS: Preparation of SnI₂

Mass of Sn metal   _____ g

Mass of I₂ plus beaker   _____ g

Mass of beaker   _____ g

Mass of I₂   _____ g

Mass of SnI₂ plus filter paper   _____ g

Mass of filter paper (furnished by instructor)   _____ g

Mass of SnI₂   _____ g

Theoretical yield of SnI₂   _____ g

Percentage yield of SnI₂   _____ %

ADVANCE STUDY ASSIGNMENT: Preparation of SnI_2

1.   The Sn^{2+} ions used to precipitate the I^- in this experiment are produced from the reaction of 4.5 g of tin metal with excess hydrochloric acid. How many moles of Sn^{2+} are formed?

_____ mol Sn^{2+}

2.   How many grams of iodine are required to produce the quantity of I^- required to react with the Sn^{2+} to produce SnI_2?

_____ g I_2

3.   What is the maximum mass of SnI_2 that can be prepared from the reaction of 4.5 g of tin with iodine?

_____ g SnI_2

4.   How might you prove that the tin iodide produced in this experiment has the formula SnI_2 rather than, say, SnI_4?

ACIDS AND BASES

EXPERIMENT

25 • pH, Its Measurement and Applications

One of the more important properties of an aqueous solution is its concentration of hydrogen ion. The H^+ or H_3O^+ ion has great effect on the solubility of many inorganic and organic species, on the nature of complex metallic cations found in solutions, and on the rates of many chemical reactions. It is important that we know how to measure the concentration of hydrogen ion and understand its effect on solution properties.

For convenience the concentration of H^+ ion is frequently expressed as the pH of the solution rather than as molarity or normality. The pH of a solution is defined by the following equation:

$$pH = -\log [H^+] \tag{1}*$$

where the logarithm is taken to the base 10. If $[H^+]$ is 1×10^{-4} mol/dm³, the pH of the solution is, by the equation, 4. If the $[H^+]$ is 5×10^{-2} M, the pH is 1.3.

Basic solutions can also be described in terms of pH. In water solutions the following equilibrium relation will always be obeyed:

$$[H^+] \times [OH^-] = K_w = 1 \times 10^{-14} \text{ at } 25°C \tag{2}$$

Since $[H^+]$ equals $[OH^-]$ in distilled water, by Equation 2, $[H^+]$ must be 1×10^{-7} M. Therefore, the pH of distilled water is 7. Solutions in which $[H^+] > [OH^-]$ are said to be acidic and will have a pH<7; if $[H^+] < [OH^-]$, the solution is basic and its pH>7. A solution with a pH of 10 will have a $[H^+]$ of 1×10^{-10} M and a $[OH^-]$ of 1×10^{-4} M.

We measure the pH of a solution experimentally in two ways. In the first of these we use a chemical called an indicator, which is sensitive to pH. These substances have colors that change over a relatively short pH range (about 2 pH units) and can, when properly chosen, be used to roughly determine the pH of a solution. Two very common indicators are litmus, usually used on paper, and phenolphthalein, the most common indicator in acid-base titrations. Litmus changes from red to blue as the pH of a solution goes from about 6 to about 8. Phenolphthalein changes from colorless to red as the pH

---

*See Experiment 19 for a discussion of the approximation made in this equation and the other equations based on equilibrium theory that appear in this experiment.

goes from 8 to 10. A given indicator is useful for determining pH only in the region in which it changes color. Indicators are available for measurement of pH in all the important ranges of acidity and basicity. Universal indicators, which contain a mixture of several indicators and show color changes over a wide pH range, are also in common use.

The other method for finding pH is with a device called a pH meter. In this device two electrodes, one of which is sensitive to $[H^+]$, are immersed in a solution. The potential between the two electrodes is related to the pH. The pH meter is designed so that the scale will directly furnish the pH of the solution. A pH meter gives much more precise measurement of pH than does a typical indicator, and is ordinarily used when an accurate determination of pH is needed.

Some acids and bases undergo substantial ionization in water, and are called *strong* because of their essentially complete ionization in reasonably dilute solutions. Other acids and bases, because of incomplete ionization (often only about 1 per cent in 0.1 M solution), are called *weak*. Hydrochloric acid, HCl, and sodium hydroxide, NaOH, are typical examples of a strong acid and a strong base respectively. Acetic acid, $HC_2H_3O_2$, sometimes abbreviated to HOAc, and ammonia, NH_3, sometimes called ammonium hydroxide, are classic examples of a weak acid and a weak base.

A weak acid will ionize according to the Law of Chemical Equilibrium:

$$HA(aq) \rightleftharpoons H^+(aq) + A^-(aq) \tag{3}$$

At equilibrium,
$$\frac{[H^+][A^-]}{[HA]} = K_a \tag{4}$$

K_a is a constant characteristic of the acid HA; in solutions containing HA, the product of concentrations in the equation will remain constant at equilibrium independent of the manner in which the solution was made up. A similar relation can be written for solutions of a weak base.

The value of the ionization constant K_a for a weak acid can be found experimentally in several ways. Perhaps the most obvious way is to simply measure the pH of a solution of the acid of known molarity. From the pH the $[H^+]$ can be calculated from (1). By (3), $[H^+]$ must be equal to $[A^-]$, since the H^+ ion is produced essentially from Reaction 3 alone. The [HA] in (4) is equal to the original molarity of HA, corrected if necessary for the amount of HA lost by (3).

Another very simple procedure involves very little calculation, is accurate, and does not even require a knowledge of the molarity of the acid. A sample of a weak acid HA, often a solid, is dissolved in water. The solution is divided into two equal parts. One part of the solution is titrated to a phenolphthalein end point with an NaOH solution, with the HA converted to A^- by the reaction:

$$OH^-(aq) + HA(aq) \rightleftharpoons A^-(aq) + H_2O \tag{5}$$

The number of moles of A^- produced equals the number of moles HA in the other part of the sample. The two solutions are then mixed, and the pH of the resultant solution is obtained. In that solution it is clear that [HA] equals $[A^-]$, so that by (4),

$$[H^+] = K_a \tag{6}$$

By this method the $[H^+]$ obtained from the pH measurement is equal to the value of the ionization constant of the acid.

Salts that can be formed by the reaction of strong acids and bases, such as NaCl, KBr, or $NaNO_3$, ionize completely but do not react with water when in solution and form neutral solutions with a pH of about 7. When dissolved in water, salts of weak acids or weak bases furnish ions that tend to react to some extent with water, producing molecules of the weak acid or base and liberating some OH^- or H^+ ion to the solution.

If HA is a weak acid, the A^- ion produced when NaA(s) is dissolved in water will react with water to some extent according to the equation

$$A^-(aq) + H_2O \rightleftharpoons HA(aq) + OH^-(aq) \qquad (7)$$

At equilibrium,

$$\frac{[HA][OH^-]}{[A^-]} = K_b; \quad K_b \ll 1 \text{ ordinarily} \qquad (8)$$

Solutions of sodium acetate, NaOAc, the salt formed by reaction of sodium hydroxide with acetic acid, will be slightly *basic*, because of the reaction of OAc^- ion with water to produce HOAc and OH^-. Because of the analogous reaction of the NH_4^+ ion with water, solutions of ammonium chloride, NH_4Cl, will be slightly *acidic*. (The reactions of OAc^- and NH_4^+ ions with water are still sometimes called hydrolysis reactions, but the term is unfortunate, even misleading. It is probably best to consider an OAc^- ion and an NH_4^+ ion as a weak base and a weak acid respectively, since that is really what they are.)

Some solutions, called buffers, are remarkably resistant to pH changes caused by the addition of an acid or base. These solutions almost always contain both the salt of a weak acid or base and the parent acid or base. The solution used in the previously mentioned procedure for finding K_a of a weak acid is an example of a buffer solution. That solution contained equal amounts of the weak acid HA and the anion A^- present in its salt. If a small amount of strong acid were added to that solution, the H^+ ion would tend to react with the A^- ion present, keeping the $[H^+]$ about where it was before the addition. Similarly a small amount of strong base added to that solution would react with the HA present, producing some A^- ion and water but not appreciably changing the $[OH^-]$. If similar amounts of acid or base were added to water the pH change could easily be several units.

To prepare a buffer that would tend to have and maintain a given pH value, we need a solution of a weak acid or base and a solution of the salt of that acid or base. Given the value of K_a of the acid and the $[H^+]$ required in an acid buffer, we can substitute into (4) to find $[A^-]/[HA]$ in the buffer. By mixing appropriate volumes of the acid and salt solutions we can easily obtain that concentration ratio and constitute the buffer. Buffers will typically have pH values that have associated $[H^+]$ ion concentrations near the K_a values for the weak acid in the buffer, since if the buffer is to be effective against addition of both acid and base, the ratio $[A^-]/[HA]$ should be as close to one as possible.

EXPERIMENTAL PROCEDURE

You may work in pairs on the first two parts of this experiment.

A. Measurement of the pH of Some Typical Acidic and Basic Solutions

Using the stock solution of 1.0 M HCl, prepare by appropriate dilutions with distilled water about 25 cm³ of each of two solutions, 0.1 M HCl and 0.01 M HCl. Measure and record the pH of the 1.0 M, and 0.1 M, and the 0.01 M solutions on the pH meter, using a 150 cm³ beaker. To each of the solutions add two drops of methyl violet indicator and record the indicator color.

Repeat these measurements, starting with the stock 1.0 M HOAc solution, but this time test the solutions (separate samples) with both methyl violet and methyl yellow indicators.

Repeat the measurements using the stock 1.0 M NH_3 solution. Use alizarin yellow as an indicator with these solutions.

B. Measurement of the pH of Some Typical Salt Solutions

Using the pH meter and 25 cm³ samples, determine and record the pH of the following stock solutions: 0.1 M NaCl, 0.1 M NaOAc, 0.1 M Na_2CO_3, 0.1 M NH_4NO_3, 0.1 M $ZnCl_2$, 0.1 M $Cu(NO_3)_2$.

C. Determination of the Dissociation Constant of a Weak Acid

In the rest of the experiment each student is to work independently.

Obtain from the stockroom a sample of an unknown acid and a buret. Measure out 100 cm³ of distilled water in a graduated cylinder and pour it into a clean 250 cm³ Erlenmeyer flask. Dissolve about half your acid sample in the water and stir thoroughly.

Pour half the solution into another 250 cm³ Erlenmeyer flask. Use the solution levels in the two flasks to decide when the volumes of solution in the two flasks are equal. Titrate the acid in one of the flasks to a phenolphthalein end point, using 0.2 M NaOH in the buret. (See Exp. 26; volume readings do not have to be taken here.) This should take less than 50 cm³ of the NaOH. Add the hydroxide solution slowly while rotating the flask. As the end point approaches, add the solution drop by drop until the solution has a permanent pink color.

Mix the neutralized solution with the acid solution in the other flask and determine the pH of the resulting half-neutralized solution. From the observed pH calculate K_a for the unknown acid.

D. The Preparation and Properties of a Buffer (optional)

Report the values of pH and K_a you obtained in C to your laboratory supervisor, who will then assign you the pH of a buffer to prepare from the rest of your solid acid.

Dissolve the remaining portion of your acid sample in 100 cm³ water, divide the solution into two equal parts, and titrate one portion with 0.2 M NaOH in a 250 cm³ flask as in C. Record the volume of NaOH solution required.

You now have a solution of the acid in one flask and a solution of its sodium salt in the other. To make the concentration of acid equal to the concentration of salt, add to the acid solution a volume of water equal to that of the 0.2 M NaOH solution used to titrate the acid portion.

Calculate on the report sheet the ratio of anion concentration to undissociated acid concentration required in the buffer. Since the anion and undissociated acid concentrations are equal in the two solutions you have prepared, in the buffer to be made by mixing these solutions,

$$\frac{\text{volume salt solution needed}}{\text{volume acid solution needed}} = \frac{[A^-]}{[HA]}$$

Using this ratio of volumes, make up at least 50 cm³ of your buffer and measure its pH. Compare the observed pH with the value you wished to obtain.

To 25 cm³ of your buffer solution add 5 drops 0.1 M NaOH and stir thoroughly. Measure the pH of the resulting solution.

To another 25 cm³ sample of your buffer add 5 drops 0.1 M HCl, mix thoroughly, and measure the pH of the mixed solution.

Measure the pH of the distilled water in the laboratory. To one 25 cm³ sample of water add 5 drops 0.1 M NaOH and to another add 5 drops 0.1 M HCl. After mixing, measure the pH of the two solutions.

OBSERVATIONS, CALCULATIONS, AND EXPLANATIONS: pH, Its Measurement and
                                                Applications

A. Measurement of the pH of Some Typical Acidic and Basic Solutions

Record the pH of the solutions and the indicator colors.

		1.0 M	0.1 M	0.01 M
HCl	pH	_____	_____	_____
	color—methyl violet	_____	_____	_____
HOAc	pH	_____	_____	_____
	color—methyl violet	_____	_____	_____
	color—methyl yellow	_____	_____	_____
NH$_3$	pH	_____	_____	_____
	color—alizarin yellow	_____	_____	_____

1. What is the per cent ionization of each of the HOAc solutions? How does the
per cent ionization of the acid vary with its concentration? Hint: First find [H$^+$] in
each solution; then compare it with [H$^+$] if the acid completely dissociated.

2. Which indicator would you suggest would be useful in determining the pH of a
5.0×10^{-4} M HCl solution? Why?

3. From the pH measurements you obtained, calculate K_a for acetic acid. (For
each solution, use [H$^+$] from Question 1, find [OAc$^-$] from Equation 3, and then use
Equation 4 to find K_a.)

B. Measurement of the pH of Some Typical Salt Solutions

Record the pH of the solutions studied.

0.1 M NaCl _____   0.1 M NH$_4$NO$_3$ _____   **197**

Continued on following page

0.1 *M* NaOAc _____   0.1 *M* ZnCl$_2$ _____

0.1 *M* Na$_2$CO$_3$ _____   0.1 *M* Cu(NO$_3$)$_2$ _____

Explain the observation of any pH in these solutions that is not within 1 pH unit of 7 by writing the net ionic equation for the reaction responsible for the pH change:

C. Determination of the Dissociation Constant of a Weak Acid

pH of half-neutralized acid solution   _____

[H$^+$] in half-neutralized solution   _____*M*

K_a for unknown acid _____   Unknown no. _____

D. Preparation and Properties of a Buffer

pH of buffer to be prepared   _____

[H$^+$] in buffer   _____*M*

$\dfrac{[A^-]}{[HA]}$ in buffer solution $=\dfrac{\text{Volume salt solution}}{\text{Volume acid solution}}$ in buffer   _____

Buffer was made by mixing _____ cm^3 salt solution with _____ cm^3 acid solution.

pH of prepared buffer   _____

Properties of prepared buffer:

pH of buffer   _____   pH of H$_2$O   _____

pH of buffer +
5 drops 0.1 *M* NaOH _____   pH of H$_2$O +
5 drops 0.1 *M* NaOH _____

pH of buffer +
5 drops 0.1 *M* HCl _____   pH of H$_2$O +
5 drops 0.1 *M* HCl _____

Conclusions:

ADVANCE STUDY ASSIGNMENT: pH, Its Measurement and Applications

1. The pH of a solution was found to be 5.0. Find $[H^+]$ in the solution.

_____M

2. A solution of Na_2CO_3 has a pH of 11.6. Find the $[H^+]$ and $[OH^-]$ in the solution. Is the solution acidic or basic? Write a net ionic equation to account for the fact that the solution is not neutral.

$[H^+] =$ _____M; $[OH^-] =$ _____M

3. A 0.2 M solution of the weak acid HB has a pH of 3.0. What is $[H^+]$ in the solution? What is $[B^-]$? [HB]? Find K_a for HB.

$[H^+] =$ _____M; $[B^-] =$ _____M; $[HB] =$ _____M; $K_a =$ _____

4. A solid acid is dissolved in water. Half the solution is titrated to a phenolphthalein end point with NaOH solution. The neutralized and acid solutions are then mixed and the pH of the resulting solution is found to be 3.8. Find K_a of the solid acid.

$K_a =$ _____

5. Acetic acid has a K_a of about 2×10^{-5}. A buffer is to be prepared with a pH of 5.0 from 0.1 M solutions of HOAc and NaOAc. How many cm^3 of 0.1 M NaOAc would have to be added to 100 cm^3 of 0.1 M HOAc to make the proper buffer solution?

_____ml   **199**

EXPERIMENT

26 • The Standardization of a Basic Solution and the Determination of the Gram Equivalent Mass of a Solid Acid

When a solution of a strong acid is mixed with a solution of a strong base, a chemical reaction occurs that can be represented by the following net ionic equation:

$$H^+(aq) + OH^-(aq) \rightarrow H_2O$$

This is called a neutralization reaction, and chemists use it extensively to change the acidic or basic properties of solutions. The equilibrium constant for the reaction is about 10^{14} at room temperature, so that the reaction can be considered to proceed completely to the right, using up whichever of the ions is present in the lesser amount and leaving the solution either acidic or basic, depending on whether H^+ or OH^- ion was in excess.

Since the reaction is essentially quantitative, it can be used to determine the concentrations of acidic or basic solutions. A frequently used procedure involves the titration of an acid with a base. In the titration, a basic solution is added from a buret to a measured volume of acid solution until the number of moles of OH^- ion added is just equal to the number of moles of H^+ ion present in the acid. At that point the volume of basic solution that has been added is measured.

Recalling the definition of the concentration term called molarity,

$$\text{molarity } M \text{ of Species } S = \frac{\text{no. moles } S \text{ in the solution}}{\text{volume of the solution (dm}^3)}$$

or, no. moles S in solution = molarity of S × volume of solution in cubic decimetres. We can see that at the end point of a titration of an acid with a base,

no. moles H^+ originally present = no. moles OH^- added

$$M_{H^+} \times V_{acid} = M_{OH^-} \times V_{base}$$

Therefore, if the molarity of either the H^+ or the OH^- ion in its solution is known, the molarity of the other ion can be found from the titration.

The equivalence point or end point in the titration is determined by using a chemical, called an indicator, that changes color at the proper point. The indicators used in acid-base titrations are weak organic acids or bases that change color when they are neutralized. One of the most common indicators is phenolphthalein, which is colorless in acid solutions but becomes red when the pH of the solution becomes 9 or higher.

When a solution of a strong acid is titrated with a solution of a strong base, the pH at the end point will be about 7. At the end point a drop of acid or base added to the solution will change its pH by several pH units, so that phenolphthalein can be used as an indicator in such titrations. If a weak acid is titrated with a strong base, the pH at the equivalence point is somewhat higher than 7, perhaps 8 or 9, and phenolphthalein is still a very satisfactory indicator. If, however, a solution of a weak base such as ammonia

201

is titrated with a strong acid, the pH will be a unit or two less than 7 at the end point, and phenolphthalein will not be as good an indicator for that titration as, for example, methyl red, whose color changes from red to yellow as the pH changes from about 4 to 6. Ordinarily, indicators will be chosen so that their color change occurs at about the pH at the equivalence point of a given acid-base titration.

In this experiment you will determine the molarity of OH^- ion in an NaOH solution by titrating that solution against a standardized solution of HCl. Since in these solutions one mole of acid in solution furnishes one mole of H^+ ion and one mole of base produces one mole of OH^- ion, $M_{HCl} = M_{H^+}$ in the acid solution, and $M_{NaOH} = M_{OH^-}$ in the basic solution. Therefore the titration will allow you to find M_{NaOH} as well as M_{OH^-}. You will note, however, that in order to be able to calculate M_{NaOH} you must know the formula of the base.

To establish the accuracy with which you standardize your NaOH solution you will use it to titrate a sample of a pure solid organic acid. By titrating a weighed sample of unknown acid with your standardized NaOH solution you can easily determine the number of moles of H^+ ion available in the sample. From the number of moles of H^+ and the mass of the sample you can calculate the number of grams of acid that would contain one mole of H^+ ion. This is called the gram equivalent mass of the acid. If one mole of the acid can produce one mole of H^+, then the mass of a mole of the acid and its gram equivalent mass are equal. If, however, the acid has three moles of available H^+ ion per mole of acid, the *GMM* is $3 \times GEM$. Since you will not be given the formula of the acid, you will be able to determine only the gram equivalent mass of the acid by titration with NaOH.

EXPERIMENTAL PROCEDURE

WEAR YOUR SAFETY GLASSES WHILE PERFORMING THIS EXPERIMENT

Note: This experiment is relatively long unless you know precisely what you are to do. Study the experiment carefully before coming to class, so that you don't have to spend a lot of time finding out what the experiment is all about.

Obtain two burets and a sample of solid unknown acid from the stockroom.

A. Standardization of NaOH Solution. Into a small graduated cylinder draw about 7 cm³ of the stock 6 *M* NaOH solution provided in the laboratory and dilute to about 400 cm³ with distilled water in a 500 cm³ Florence flask. Stopper the flask tightly and mix the solution thoroughly at intervals over a period of at least 15 min before using the solution.

Draw into a clean *dry* 125 cm³ Erlenmeyer flask about 75 cm³ of standardized HCl solution (about 0.1 *M*) from the stock solution on the reagent shelf. This amount should provide all the standard acid you will need, so do not waste it.

Prepare for the titration using the procedure given in Experiment 22. Clean the two burets and rinse with distilled water. Then rinse the first buret three times with a little of the HCl solution. Fill the buret with HCl; open the stopcock momentarily to fill the tip. Proceed to clean and fill the other buret with your NaOH solution in a similar manner. Carefully label the two burets. Check to see that your burets do not leak and that there are no air bubbles in either buret tip. Read the levels in both burets to 0.02 cm³.

Draw about 25 cm³ of the HCl solution from the buret into a clean 250 cm³ Erlenmeyer flask; add to the flask about 25 cm³ of distilled H_2O and 2 or 3 drops of phenolphthalein indicator solution. Place a white sheet of paper under the flask to aid in the detection of any color change. Add the NaOH solution intermittently from its buret to the solution in the flask, noting the pink phenolphthalein color that appears and disappears as the drops hit the liquid and are mixed with it. Swirl the liquid in the flask gently and continuously as you add the NaOH solution. When the pink color begins to persist, slow down the rate of addition of NaOH. In the final stages of the titration add the NaOH drop by drop until the entire solution just turns a pale pink color that will persist for about 30 s. If you go past the end point and obtain a red solution, add a few drops of the HCl solution to remove the color, and then add NaOH a drop at a time until the pink color persists. Carefully record the final readings on the HCl and NaOH burets.

To the 250 cm³ Erlenmeyer flask containing the titrated solution, add about 10 cm³ more of the standard HCl solution. Titrate this as before with the NaOH to an end point, and carefully record both buret readings once again. To this solution add about 10 cm³ more HCl and titrate a third time with NaOH.

You have now completed three titrations, with *total* HCl volumes of about 25, 35, and 45 cm³. Find the ratio $V_{NaOH}:V_{HCl}$ at the end point of each of the titrations, using *total* volumes of each reagent reacted up to that end point. At least two of these volume ratios should agree to within one per cent. If they do, proceed to the next part of the experiment. If they do not, repeat these titrations until two volume ratios do agree.

B.   Determination of the Gram Equivalent Mass of an Acid.   Weigh the vial containing your solid acid on the analytical balance. Carefully pour out about half the sample into a clean but not necessarily dry 250 cm³ Erlenmeyer flask. Weigh the vial accurately. Add about 50 cm³ of distilled water and 2 or 3 drops of phenolphthalein to the flask. The acid may be relatively insoluble, so don't worry if it doesn't all dissolve.

Fill your NaOH buret with your (now standardized) NaOH solution and read the level accurately.

Titrate the acid solution as before. As the acid is neutralized by the NaOH, it will tend to dissolve in the solution. If your unknown is so insoluble that the phenolphthalein color appears before all the solid dissolves, add 25 cm³ of ethanol to the solution to increase the solubility. Record the NaOH buret reading at the end point.

Pour the rest of your acid sample into a 250 cm³ Erlenmeyer flask and weigh the vial accurately. Titrate the acid as before with NaOH solution. If you go past the end point in these titrations, it is possible, though more complicated in calculations, to back-titrate with a little of the standard HCl solution. Measure the volume of HCl used and subtract the number of moles HCl in that volume from the number of moles NaOH used in the titration. The difference will equal the number of moles NaOH used to neutralize the acid sample.

DATA: **Standardization of a Basic Solution.**
Determination of the GEM of an Acid

A. Standardization of the NaOH Solution

	Trial 1	Trial 2	Trial 3
Final reading HCl buret	_____ cm^3	_____ cm^3	_____ cm^3
Initial reading HCl buret	_____ cm^3		
Final reading NaOH buret	_____ cm^3	_____ cm^3	_____ cm^3
Initial reading NaOH buret	_____ cm^3		

B. Gram Equivalent Mass of Unknown Acid

Mass of vial plus contents	_____ g
Mass of vial plus contents less sample 1	_____ g
Mass less sample 2	_____ g

	Trial 1	Trial 2
Final reading NaOH buret	_____ cm^3	_____ cm^3
Initial reading NaOH buret	_____ cm^3	_____ cm^3

CALCULATIONS

A. Standardization of NaOH Solution

	Trial 1	Trial 2	Trial 3
Total volume HCl	_____ cm^3	_____ cm^3	_____ cm^3
Total volume NaOH	_____ cm^3	_____ cm^3	_____ cm^3

Continued on following page

Volume ratio: V_{HCl}/V_{NaOH}
(should agree within
1 per cent)

_____   _____   _____

Molarity M_A of
standardized HCl   _____ M

No. moles acid $= \dfrac{V_A M_A}{1000} =$ no. moles base $= \dfrac{V_B M_B}{1000}$ (where V_A, V_B are in ml)

Molarity M_B of NaOH
solution   _____ M _____ M _____ M

Average molarity of NaOH solution   _____ M

B. Gram Equivalent Mass of an Unknown Acid

Mass of sample   _____ g   _____ g

Volume NaOH used   _____ cm^3 _____ cm^3

No. moles NaOH $= \dfrac{V_{NaOH} M_{NaOH}}{1000}$ _____   _____

No. moles OH$^-$ = no. moles NaOH   _____   _____

No. moles H$^+$ in sample   _____   _____

GEM $= \dfrac{\text{grams acid}}{\text{moles H}^+}$ _____ g   _____ g

Unknown No.   _____

Name _____ **Section** _____

ADVANCE STUDY ASSIGNMENT: Gram Equivalent Mass of an Unknown Acid

1. If 7.0 cm³ of 6.0 M NaOH are diluted with water to a volume of 400 cm³, what is the molarity of the resulting solution?

_____ M

2. In an acid-base titration, 22.81 cm³ of an NaOH solution were required to neutralize 26.18 cm³ of a 0.1121 M HCl solution. What is the molarity of the NaOH solution?

_____ M

3. A gram equivalent mass, or an equivalent, of an acid contains one mole of acid hydrogen. How many equivalents of acid would there be in 0.3 mol HNO_3? In 0.5 mol H_2SO_4?

_____   _____

4. A 0.2861 g sample of an unknown solid acid required 32.63 cm³ of 0.1045 M NaOH for neutralization to a phenolphthalein end point. What is the gram equivalent mass of the acid?

_____ g

5. Why is it not possible to find the molecular mass of an unknown acid by the method used in this experiment? How might the molecular mass of the acid be found?

27 • Determination of the Ionization Constant of a Weak Acid

When a weak acid HA is dissolved in water, it dissociates to some extent according to the equation

$$HA(aq) \rightleftharpoons H^+(aq) + A^-(aq) \tag{1}$$

The dissociation reaction for the acid obeys the law of chemical equilibrium, so that in the solution at equilibrium,

$$\frac{[H^+][A^-]}{[HA]} = K_a \tag{2}$$

where K_a is a constant independent of the amount of HA originally dissolved. K_a is called the *ionization constant* of the acid; it is a characteristic of the acid and will have a fixed value at a given temperature. If a solution is made by dissolving arbitrary amounts of H^+ ion, A^- ion, and HA, Reaction 1 will occur to the right or the left until, at equilibrium, Equation 2 is satisfied. If we then dilute the solution with water, thus disturbing the equilibrium, Reaction 1 will again occur, changing the concentrations of all species until the equilibrium condition in Equation 2 is once again obeyed.

To determine the value of the ionization constant K_a for a weak acid, we must in general find the values of $[H^+]$, $[A^-]$, and $[HA]$ in at least one solution containing those species at equilibrium; substitution of those values into Equation 2 yields a value for K_a that will be applicable to any other solution of the acid at the same temperature.

If the equilibrium solution were initially made up by dissolving a known amount of pure HA in a measured volume of solution, then, by Equation 1, it would be true that

$$[H^+] = [A^-] \quad \text{and} \quad [HA] = [HA]_0 - [H^+] \tag{3}$$

where $[HA]_0$ is the concentration HA would have had if there were no dissociation at all; you will recognize $[HA]_0$ as being equal to the molarity M of the solution. Under these somewhat restricted conditions K_a could be calculated from one direct measurement of $[H^+]$ and the relations in Equation 3. In this experiment we will determine K_a for an unknown weak acid under these conditions.

You will recall that in reporting $[H^+]$ in solutions one frequently employs an alternate notation involving a quantity called pH. The pH of a solution is defined by the equation

$$pH = -\log[H^+] \tag{4}$$

By Equation 4, the pH of a solution in which $[H^+]$ equals $0.01\ M$ will be equal to 2.0; the $[H^+]$ in a solution having a pH of 4.0 is equal to $1 \times 10^{-4}\ M$. The two ways of expressing hydrogen ion concentration are both commonly used, and it is important that you be familiar with them.

In order to find the pH or $[H^+]$ in a solution we ordinarily use one of two methods. The first employs a device called a pH meter, which automatically determines the pH of a solution by measuring an electric potential that is related to the pH; this device is used in Experiment 25. In this experiment we will use another method, involving substances called acid-base indicators, which are sensitive to the $[H^+]$ in a solution. In Experiment 26, phenolphthalein, a very common acid-base indicator, was used to determine the end point of an acid-base titration. Phenolphthalein is colorless in acid solution but turns pink when the pH of a solution exceeds about 9.

Most acid-base indicators, unlike phenolphthalein, change color over a rather narrow pH range and can be used to establish the pH of an unknown solution. These indicators are themselves weak organic acids that have characteristic colors in both the molecular and dissociated forms. Consider the dissociation of the weak indicator acid HIn:

$$HIn(aq) \rightleftharpoons H^+(aq) + In^-(aq) \qquad K_{Ind} = \frac{[H^+][In^-]}{[HIn]} \qquad (5)$$

If a small amount of HIn is added to an acid solution the $[H^+]$ will be fixed by the solution, and so

$$\frac{[In^-]}{[HIn]} = \frac{K_{Ind}}{[H^+]}$$

If the color of the In^- ion is yellow and the color of HIn is red, as is the case with several indicators, the actual color of a solution containing the indicator will depend on K_{Ind} and $[H^+]$. If $[H^+] \ll K_{Ind}$, then $[In^-]/[HIn]$ will be very large, and the solution will be yellow. If, on the other hand, $[H^+] \gg K_{Ind}$, then $[In^-]/[HIn]$ will be very small and the indicator will be primarily in the red HIn form. If $[H^+]$ is about equal to K_{Ind}, both forms will be present in about equal concentrations, and the indicator will appear orange in the solution. Since it is difficult to detect less than about 10 per cent of one form of the indicator in the presence of 90 per cent of the other form, most indicators will change effectively from one color to the other for a change in $[H^+]$ by a factor of about 100, or 2 pH units. Over the range where the color is changing, the indicator is useful; outside that range it is not, and another indicator must be used.

Given an indicator that in its useful range produces a given color in an unknown acid solution, the $[H^+]$ or the pH of that solution can be determined by preparing a known solution of such a $[H^+]$ that the indicator has the same color in both the known and the unknown solutions. Under those conditions, it is clear that

$$[H^+]_{known} = [H^+]_{unknown} \quad \text{and} \quad pH_{known} = pH_{unknown} \qquad (6)$$

In this experiment you will determine the pH of several acid solutions by means of indicators. You will be given a solution of an unknown acid of known molarity. The experiment will involve your finding the pH of that solution and three other solutions obtained from it by dilution with water.

In the laboratory you will be provided with a standard solution of 0.1 M HCl. Since HCl is a strong acid, unlike your unknown, it can be considered to be completely ionized, so that in your standard, $[H^+]$ will equal 0.1 M and the pH will be 1.0. You can prepare other solutions of known pH from this standard solution by diluting it with water. These solutions will serve as the references on which you can base your pH measurements for your unknown solutions.

There are many acid-base indicators, but in this experiment we will be concerned with those which are useful in the pH range from about 2 to 5. Those indicators which are active in this range are listed in the table with their pH range and the color change that occurs as the pH is increased.

For each solution of your acid unknown you will first determine which indicator is useful for the pH measurement, in that with your unknown it gives a color somewhere between the limiting values. By comparing the color of your unknown with that of your standard solutions with the same indicator, you can easily fix the pH of the unknown to within about one pH unit. Then, by diluting one of your standard solutions with

Indicator	pH Range	Color Change
Methyl violet	0 to 2	yellow to violet
Meta cresol purple or thymol blue	1 to 3	red to yellow
Methyl yellow	2.9 to 4	red to yellow
Methyl orange	3 to 5	red to yellow
Bromphenol blue	3 to 5	yellow to blue
Bromcresol green	4 to 6	yellow to blue

water, you can prepare a known solution, with a pH you can calculate, in which the indicator produces a color that matches exactly the color of the unknown. The pH of the unknown is then determined by Equation 6. With care your pH measurements can be accurate to ± 0.2 pH unit.

Having found the pH and hence $[H^+]$ in each solution it will be possible to use the relations in Equation 3 and substitute into Equation 2 to find K_a for the unknown acid. Under the conditions of the experiment,

$$K_a = \frac{[H^+]^2}{[HA]} \tag{7}$$

We shall actually make the calculation in a slightly different manner in order to minimize experimental errors. First, let us rewrite Equation 7, taking the logarithms of both sides,

$$\log K_a = 2 \log [H^+] - \log [HA]$$

Recalling that

$$pH = -\log [H^+]$$

we see that

$$\log K_a = -2\,pH - \log [HA]$$

or

$$2\,pH = -\log K_a - \log [HA] \tag{8}$$

If you make a graph of 2 pH of the solutions against their $\log [HA]$, the points on the graph should lie on a straight line of slope equal to -1; the line will cut the pH co-ordinate at $-\log K_a$ when $\log [HA]$ equals zero. K_a will be found from $\log K_a$ as obtained from a graph prepared in this way.

EXPERIMENTAL PROCEDURE

WEAR YOUR SAFETY GLASSES WHILE PERFORMING THIS EXPERIMENT

Obtain a sample of unknown acid of known molarity from the stockroom.

Draw 50 cm³ 0.1 M HCl from the stock bottle into a clean 125 cm³ Erlenmeyer flask; label the flask with the pH of the acid. Using a pipet you have cleaned and rinsed twice with one or two ml of the solution, pipet 5.0 cm³ of this solution into a second 125 cm³ flask. Add 45 cm³ of distilled water from your graduated cylinder to the acid in the flask and mix thoroughly; label this flask with its proper pH. Rinse your pipet twice with the acid in this flask, and then pipet 5.0 cm³ of this 0.01 M acid into another clean 125 cm³ flask. Dilute this acid with 45 cm³ distilled water, mix thoroughly, and again label it with its pH. Repeat this procedure until you have five solutions, with known pH values of 1.0, 2.0, 3.0, 4.0, and 5.0.

Test 1 cm³ samples of your unknown with a drop or two of the available indicators, noting the colors you obtain and so determining roughly the pH of your unknown. Then test 1 cm³ samples of your known solutions having a pH near that of your unknown with the indicator or indicators that appear to be useful in the pH range of your unknown. By this procedure you should be able to determine that the pH of your unknown lies in a given one-unit pH interval, say between a pH of 3 and a pH of 4. By mixing appropriate volumes of the more acidic known solution and water, as measured in your graduated cylinder, prepare a solution which, with a useful indicator, has exactly the same color as your unknown. (If the volume of this solution is significantly larger than that of your unknown, add another drop or two of indicator to the solution.) Calculate the pH of the known solution from its composition; since the indicator colors match, the pH of your unknown will be the same as that of the known.

Rinse your pipet with your unknown twice, and then pipet 5.0 cm³ of the unknown into a clean 125 cm³ flask and dilute with 45 cm³ of distilled water. Mix the solution well and determine its pH by the method you used with the original sample. Carry out two more tenfold dilutions of your unknown and determine the pH of each of these solutions.

DATA AND CALCULATIONS: **Determination of the Ionization Constant of a Weak Acid**

Unknown no. _____   Initial molarity of unknown _____ M

I. Color of Unknown in Presence of Indicators

Methyl violet	Meta cresol purple	Methyl yellow	Methyl orange	Bromphenol blue	Bromcresol green
_____	_____	_____	_____	_____	_____

Approximate pH of unknown _____

Indicator selected for final determination _____

Composition of known solution matching color of unknown:

_____ cm^3 of pH _____ solution   (A)

plus _____ cm^3 of water

No. of moles H$^+$ in A $= \dfrac{\text{Volume}_A \times [\text{H}^+]_A}{1000} =$ _____ moles $= n$

Total volume of matching solution $=$ _____ cm^3 $= V$

$[\text{H}^+]$ in matching solution $= \dfrac{1000\,n}{V} =$ _____ M

pH of matching solution $=$ pH of unknown solution $=$ _____

Continued on following page   **213**

II.   First Dilution of Unknown

Approximate pH _____

Indicator used for
final determination _____

Actual pH of unknown _____

Molarity of unknown _____ *M*

Calculation of pH of matching solution:

III.   Second Dilution of Unknown

Approximate pH _____

Indicator used for
final determination_____

Actual pH of unknown _____

Molarity of unknown _____ *M*

Calculation of pH of matching solution:

IV.   Third Dilution of Unknown

Approximate pH _____

Indicator used for
final determination_____

Actual pH of unknown _____

Molarity of unknown _____ *M*

Calculation of pH of matching solution:

Calculation of K_a for unknown acid: $2\,\mathrm{pH} = -\log K_a - \log [\mathrm{HA}]$ (8)

Equation 8 will be used in finding the value of K_a. To use this equation it is perhaps easiest to first complete the table below for the four unknown acid solutions studied:

Solution	Measured pH	Molarity M of Unknown Acid	log M
I	_____	_____	_____
II	_____	_____	_____
III	_____	_____	_____
IV	_____	_____	_____

In Equation 8, log [HA] refers to the molarity of the *undissociated* acid, not to the overall molarity M of the acid solution. $M = [\mathrm{HA}]_0$; $[\mathrm{HA}] = [\mathrm{HA}]_0 - [\mathrm{H}^+]$. Since $[\mathrm{H}^+]$ is $\ll [\mathrm{HA}]_0$, it is approximately true that M equals [HA]; the experimental error in the determination of pH is much larger than the error in this approximation. In Equation 8,

Continued on following page

log [HA] may for these reasons be taken to equal log M. Therefore, within experimental error,

$$2\,\mathrm{pH} = -\log K_a - \log M \qquad (8')$$

Make a graph on which the ordinate is equal to $2\,\mathrm{pH}$ and the abscissa is $\log M$. Plot the points in the table as determined for each solution. Draw a straight line through the points in such a way as to minimize the distances from the points to the line, and extrapolate the line to find the value of $2\,\mathrm{pH}$ at $\log M$ equals zero. By Equation 8, this value will equal $(-\log K_a)$ for the weak acid.

Value of $(-\log K_a)$ obtained from the graph _____

K_a for the unknown acid _____

DETERMINATION OF THE IONIZATION CONSTANT
OF A WEAK ACID (Data and Calculations)

ADVANCE STUDY ASSIGNMENT: **Determination of the Ionization**
                                Constant of a Weak Acid

1.   A $0.1\,M$ solution of a weak acid HA has a pH of 3.3. What is the $[H^+]$ in the solution? What is $[A^-]$? $[HA]$? Find the ionization constant K_a for the acid.

$[H^+] =$ _____M; $[A^-] =$ _____M; $[HA] =$ _____M; $K_a =$ _____

2.   A $0.5\,M$ solution of an acid is tested with some of the indicators used in this experiment. The colors observed were: in methyl violet, violet; in meta cresol purple, yellow; in methyl orange, yellow; in bromcresol green, green. What is the approximate pH of the acid solution? Which indicator would be used to find the pH precisely?

_____

_____

3.   The color of the acid solution in Problem 2 with bromcresol green indicator matches very closely the color of bromcresol green in a solution made by mixing $2.0\ cm^3$ of 0.0001 M HCl with $8.0\ cm^3$ of water. What is $[H^+]$ in the matching solution? What is the pH of the acid solution? What is the value of its ionization constant?

$[H^+] =$ _____

pH $=$ _____

$K_a =$ _____

COMPLEX IONS

EXPERIMENT

28 • Relative Stabilities of Complex Ions and Precipitates Prepared from Solutions of Copper(II)

In aqueous solution, typical cations, particularly those produced from atoms of the transition metals, do not exist as free ions but rather consist of the metal ion in combination with some water molecules. Such cations are called complex ions. The water molecules, usually 2, 4, or 6 in number, are bound chemically to the metallic cation, but often rather loosely, with the electrons in the chemical bonds being furnished by one of the unshared electron pairs from the oxygen atoms in the H_2O molecules. Zinc ion in aqueous solution may exist as $Zn(H_2O)_4^{2+}$, with the water molecules arranged tetrahedrally around the central metal ion.

If a hydrated cation such as $Zn(H_2O)_4^{2+}$ is mixed with other species that can, like water, form coordinate covalent bonds with Zn^{2+}, those species, called ligands, may displace one or more H_2O molecules and form other complex ions containing the new ligands. For instance, NH_3, a reasonably good coordinating species, may replace H_2O from the hydrated zinc ion, $Zn(H_2O)_4^{2+}$, to form $Zn(H_2O)_3NH_3^{2+}$, $Zn(H_2O)_2(NH_3)_2^{2+}$, $Zn(H_2O)(NH_3)_3^{2+}$, or $Zn(NH_3)_4^{2+}$. At moderate concentrations of NH_3, essentially all the H_2O molecules around the zinc ion are replaced by NH_3 molecules, forming the zinc ammonia complex ion.

Coordinating ligands differ in their tendency to form bonds with metallic cations, so that in a solution containing a given cation and several possible ligands, an equilibrium will develop in which most of the cations are coordinated with those ligands with which they form the most stable bonds. There are many kinds of ligands, but they all share the common property that they possess an unshared pair of electrons which they can donate to form a coordinate covalent bond with a metal ion. In addition to H_2O and NH_3, other uncharged coordinating species include CO and ethylenediamine; some common anions that can form complexes include OH^-, Cl^-, CN^-, SCN^-, and $S_2O_3^{2-}$.

As you know, when solutions containing metallic cations are mixed with other solutions containing ions, precipitates are sometimes formed. When a solution of 0.1 M zinc nitrate is mixed with a little 1 M NH_3 solution, a precipitate forms and then dissolves in excess ammonia. The formation of the precipitate helps us to understand what is occurring as NH_3 is added. The precipitate is hydrous zinc hydroxide, formed by reaction of the hydrated zinc ion with the small amount of hydroxide ion present in the NH_3 solution. The fact that this reaction occurs means that even at very low OH^- ion concentration $Zn(OH)_2(H_2O)_2(s)$ is a more stable species than $Zn(H_2O)_4^{2+}$ ion.

Addition of more NH_3 causes the solid to redissolve. The zinc species then in solution cannot be the hydrated zinc ion. (Why?) It must be some other complex ion, and is, indeed, the $Zn(NH_3)_4^{2+}$ ion. The implication of this reaction is that the $Zn(NH_3)_4^{2+}$ ion is also more stable in NH_3 solution than is the hydrated zinc ion. To deduce in addition that the zinc ammonia complex ion is also more stable in general than $Zn(OH)_2(H_2O)_2(s)$ is not warranted, since under the conditions in the solution $[NH_3]$ is much larger than $[OH^-]$, and given a higher concentration of hydroxide ion, the solid hydrous zinc hydroxide might possibly precipitate even in the presence of substantial concentrations of NH_3.

To resolve this question, you might proceed to add a little 1 M NaOH solution to the solution containing the $Zn(NH_3)_4^{2+}$ ion. If you do this you will find that $Zn(OH)_2(H_2O)_2(s)$ does indeed precipitate, but that on addition of more 1 M NaOH, it too redissolves.

We can conclude from these observations that $Zn(OH)_2(H_2O)_2(s)$ is more stable than $Zn(NH_3)_4^{2+}$ in solutions in which the ligand concentrations (OH^- and NH_3) are roughly equal, but also that there is yet another species formed in the presence of high OH^- ion concentrations that is even more stable than the solid zinc hydroxide. That species we can identify chemically as $Zn(OH)_4^{2-}$.

The zinc species that will be present in a system depends, as we have just seen, on the conditions in the system. We cannot say in general that one species will be more stable than another; the stability of a given species depends in large measure on the kinds and concentrations of other species that are also present with it.

Another way of looking at the matter of stability is through equilibrium theory. Each of the zinc species we have mentioned can be formed in a reaction between the hydrated zinc ion and a complexing or precipitating ligand; each reaction will have an associated equilibrium constant, which we might call a formation constant for that species. The pertinent formation reactions and their constants for the zinc species we have been considering are listed here:

$$Zn(H_2O)_4^{2+}(aq) + 4\,NH_3(aq) \rightleftharpoons Zn(NH_3)_4^{2+}(aq) + 4\,H_2O \qquad K_1 = 3 \times 10^9 \qquad (1)$$

$$Zn(H_2O)_4^{2+}(aq) + 2\,OH^-(aq) \rightleftharpoons Zn(OH)_2(H_2O)_2(s) + 2\,H_2O \qquad K_2 = 2 \times 10^{16} \qquad (2)$$

$$Zn(H_2O)_4^{2+}(aq) + 4\,OH^-(aq) \rightleftharpoons Zn(OH)_4^{2-}(aq) + 4\,H_2O \qquad K_3 = 3 \times 10^{15} \qquad (3)$$

The formation constants for these reactions do not involve $[H_2O]$ terms, which are essentially constant in aqueous systems and are included in the magnitude of K in each case. The large size of each formation constant indicates that the tendency for the hydrated zinc ion to react with the ligands listed is very high.

In terms of these data, let us compare the stability of the $Zn(NH_3)_4^{2+}$ complex ion with that of the $Zn(OH)_4^{2-}$ complex ion. This is most readily done by considering the reaction

$$Zn(NH_3)_4^{2+}(aq) + 4\,OH^-(aq) \rightleftharpoons Zn(OH)_4^{2-}(aq) + 4\,NH_3(aq)$$

$$K = \frac{[Zn(OH)_4^{2-}]\,[NH_3]^4}{[Zn(NH_3)_4^{2+}]\,[OH^-]^4} \qquad (4)$$

Since $K_1 = \dfrac{[Zn(NH_3)_4^{2+}]}{[Zn(H_2O)_4^{2+}]\,[NH_3]^4}$ and $K_3 = \dfrac{[Zn(OH)_4^{2-}]}{[Zn(H_2O)_4^{2+}]\,[OH^-]^4}$

and since all three equilibria must be satisfied in the solution, it is clear that

$$K = \frac{K_3}{K_1} = \frac{3 \times 10^{15}}{3 \times 10^9} = 10^6$$

From the expression for K in Equation 4, we can calculate that in a solution in which the NH_3 and OH^- ligand concentrations are about equal,

$$\frac{[Zn(OH)_4^{2-}]}{[Zn(NH_3)_4^{2+}]} = 10^6$$

which means that the zinc is primarily in the form of the hydroxide complex. But that is exactly what we discovered by treating the hydrated zinc ion first with ammonia and then with an equivalent amount of hydroxide ion.

Starting now from the experimental behavior of the zinc ion, we can conclude that since the hydroxide complex ion is the one that exists when zinc ion is exposed to equal concentrations of ammonia and hydroxide ion, the hydroxide complex is more stable under those conditions, *and* the equilibrium constant for the formation of the hydroxide complex is larger than the constant for the formation of the ammonia complex. By determining, then, which species is present when a cation is in the presence of equal ligand concentrations, we can speak meaningfully of stability under such conditions and can rank the formation constants for the possible complex ions, and indeed for precipitates, in order of their increasing magnitudes.

In this experiment you will carry out formation reactions for a group of complex ions and precipitates involving the Cu^{2+} ion. You can make these species by mixing a solution of $Cu(NO_3)_2$ with solutions containing NH_3 or anions, which may form either precipitates or complex ions by reaction with $Cu(H_2O)_4^{2+}$, the cation present in aqueous solutions of copper(II) nitrate. By examining whether the precipitates or complex ions formed by the reaction of hydrated copper(II) ion with a given species can, on addition of a second ligand, be dissolved or transformed to another species, you will be able to rank the relative stabilities of the precipitates and complex ions made from Cu^{2+} with respect to one another, and thus rank the equilibrium formation constants for each species in order of increasing magnitude. The species to be reacted with Cu^{2+} ion in aqueous solution are NH_3, Cl^-, OH^-, CO_3^{2-}, $C_2O_4^{2-}$, S^{2-}, NO_2^-, and PO_4^{3-}. In each case the test for relative stability will be made in the presence of essentially equal concentrations of the two ligands. When you have completed your ranking of the known species you will be given an unknown species to incorporate into your list.

EXPERIMENTAL PROCEDURE

Obtain from the stockroom an unknown and enough medium size test tubes so that you have a total of eight.

Add about 2 cm³ 0.1 M $Cu(NO_3)_2$ solution to each of the test tubes.

To each of the test tubes add about 2 cm³ 1 M NH_3 solution, drop by drop. Report your observation in the space corresponding to NH_3–NH_3 in the table on your data page. Report the color of any precipitate that forms, and if the precipitate redissolves in excess NH_3, report the color of the resulting solution. At the bottom of the space indicate the formula of the species present in excess NH_3. If a solution is present, the Cu^{2+} ion is present in a complex, with four ligand NH_3 molecules coordinated to it. If a precipitate is present, the species is neutral and in the case of NH_3 would be a hydrous copper hydroxide, $Cu(H_2O)_2(OH)_2(s)$.

To test the stability of the copper species present in excess NH_3 relative to those possibly present with other precipitating or coordinating species, add 2 cm³ of 1 M solutions of the anions in the horizontal row in the table to the test tubes you have just prepared, one solution to a test tube. Note in the appropriate spaces in the table any changes which occur. A change in color of the solution or the formation of a new precipitate implies that a reaction has occurred between the added ligand or precipitating anion and the species originally present. Recognizing that the formulas of precipitates are neutral and that copper(II) ion typically exhibits a coordination number of four, write the formula of the species present in a system containing both excess NH_3 and the added species. If a new species is formed on addition of the second reagent, is that species more or less stable than the one originally present?

Repeat the above series of experiments, using 1 M Cl^- as the species originally added to the $Cu(NO_3)_2$ solution. In each case record the color of any precipitates or solutions formed on addition of the reagents in the horizontal row, and the formulas of the species present when an excess of both Cl^- ion and the added species is present in the solution. Since these reactions are reversible, it is not necessary to retest Cl^- with NH_3, since the same results would be obtained as when the NH_3 solution was tested with Cl^- solution.

Repeat the series of experiments for each of the anions in the vertical row in the table, omitting those tests where decisions as to relative stabilities are already clear. Where both ligands produce precipitates it may be helpful to check the effect of the addition of the other ligand to those precipitates. When complete, your table should have at least 36 entries.

Examine your table and decide on the relative stabilities of all species you observed to be present in all the spaces of the table. There should be eight such species; rank them as best you can in order of increasing stability. There is only one correct ranking, and you should be prepared to defend your choices. Although we did not in general prepare the species by direct reaction of $Cu(H_2O)_4^{2+}$ with the added ligand or precipitating anion, the equilibrium formation constants for those species for the direct reactions will have magnitudes that increase in the same order as the relative stabilities of the species you have established.

When you are satisfied that your ranking order is correct, carry out the necessary tests on your unknown to determine its proper position in the list. Your unknown may be one of the species you have already observed, or it may be a different species, present in excess of its ligand or precipitating anion.

DATA AND OBSERVATIONS: **Relative Stabilities of Complex Ions**
                              and Precipitates Containing Cu(II)

Table of Observations

	NH_3	Cl^-	OH^-	CO_3^{2-}	$C_2O_4^{2-}$	S^{2-}	NO_2^-	PO_4^{3-}
NH_3								
Cl^-								
OH^-								
CO_3^{2-}								
$C_2O_4^{2-}$								
S^{2-}								
NO_2^-								
PO_4^{3-}								
Un-known								

Continued on following page

Determination of Relative Stabilities

In each row of the table you can compare the stabilities of species involving the reagent in the horizontal row with those of the species containing the reagent initially added. In the first row of the table, the copper(II)–NH_3 species can be seen to be more stable than some of the species obtained by addition of the other reagents, and less stable than others. Examining each row, make a list of all the complex ions and precipitates you have in the table in order of increasing stability and formation constant.

Reasons

Lowest _____   _____

_____   _____

_____   _____

_____   _____

_____   _____

_____   _____

Highest _____   _____

Stability of Unknown

Indicate the position your unknown would occupy in the above list.

Reasons:

Unknown no. _____

ADVANCE STUDY ASSIGNMENT: **Stabilities of Complex Ions and Precipitates**

1. For the formation of AgBr from ions in solution,

$$Ag^+(aq) + Br^-(aq) \rightleftharpoons AgBr(s) \qquad K_1 = 1 \times 10^{13}$$

For the formation of AgI from ions in solution,

$$Ag^+(aq) + I^-(aq) \rightleftharpoons AgI(s) \qquad K_2 = 1 \times 10^{16}$$

Find the equilibrium constant for the reaction

$$AgBr(s) + I^-(aq) \rightleftharpoons AgI(s) + Br^-(aq)$$

_____

What would be the silver-containing species present at equilibrium when 2 cm³ 1 M NaBr and 2 cm³ 1 M NaI are added in succession to 2 cm³ 0.1 M AgNO₃?

_____

2. Copper(II) nitrate solution yields a precipitate when treated with a reagent containing A^-. The precipitate dissolves both in a reagent containing B^- and in a reagent containing C^-.

If the reagent containing C^- is added to a solution made by mixing copper(II) nitrate with a solution of B^-, the color of the solution undergoes a change. However, when B^- is added to a solution made by mixing copper nitrate with a solution of C^-, no change of color is observed.

Write the formulas of the copper-containing species present when copper(II) nitrate is mixed with a solution containing A^-, B^-, or C^-. Rank these species in order of increasing equilibrium formation constants.

_____   _____   _____

Increasing → _____   _____   _____

3. One way of considering relative stabilities of species of the sort studied in this experiment is to note that the more stable of two possible species will be in equilibrium with the lower concentration of hydrated cation. Calculate $[Zn(H_2O)_4^{2+}]$ in equilibrium with $Zn(NH_3)_4^{2+}$ in a solution in which $[NH_3]$ is 1 M and $[Zn(NH_3)_4^{2+}]$ is 0.1 M. Use data on page 220.

_____ M **225**

29 • Determination of the Hardness of Water

One of the factors that establishes the quality of a water supply is its degree of hardness. The hardness of water is defined in terms of its content of calcium and magnesium ions. Since the analysis does not distinguish between Ca^{2+} and Mg^{2+}, and since most hardness is caused by carbonate deposits in the earth, hardness is usually reported as total parts per million calcium carbonate by weight. A water supply with a hardness of one hundred parts per million would contain the equivalent of 100 g of $CaCO_3$ in 1 million grams of water or 0.1 gram in 1 dm^3 of water. In the days when soap was more commonly used for washing clothes, and when people bathed in tubs instead of using showers, water hardness was more often directly observed than it is now, since Ca^{2+} and Mg^{2+} form insoluble salts with soaps and make a scum that sticks to clothes or to the bath tub. Detergents have the distinct advantage of being effective in hard water, and this is really what allowed them to displace soaps for laundry purposes.

Water hardness can be readily determined by titration with the chelating agent EDTA (ethylenediaminetetraacetic acid). This reagent is a weak acid that can lose four protons on complete neutralization; its structural formula is

$$
\begin{array}{ccc}
HOOC-CH_2 & & CH_2-COOH \\
& \diagdown \quad \diagup & \\
& N-CH_2-CH_2-N & \\
& \diagup \quad \diagdown & \\
HOOC-CH_2 & & CH_2-COOH
\end{array}
$$

The four acid sites and the two nitrogen atoms all contain unshared electron pairs, so that a single EDTA ion can form a complex with up to six sites on a given cation. The complex is typically quite stable, and the conditions of its formation can ordinarily be controlled so that it contains EDTA and the metal ion in a 1:1 mol ratio. In a titration to establish the concentration of a metal ion, the EDTA which is added combines quantitatively with the cation to form the complex. The end point occurs when essentially all of the cation has reacted.

In this experiment we will standardize a solution of EDTA by titration against a standard solution made from calcium carbonate, $CaCO_3$. We will then use the EDTA solution to determine the hardness of an unknown water sample. Since both EDTA and Ca^{2+} are colorless, it is necessary to use a rather special indicator to detect the end point of the titration. The indicator we will employ is called Eriochrome Black T, which forms a rather stable wine-red complex, $MgIn^-$, with the magnesium ion. A tiny amount of this complex will be present in the solution during the titration. As EDTA is added, it will complex free Ca^{2+} and Mg^{2+} ions, leaving the $MgIn^-$ complex alone until essentially all of the calcium and magnesium has been converted to chelates. At this point EDTA concentration will increase sufficiently to displace Mg^{2+} from the indicator complex; the indicator reverts to an acid form, which is sky blue, and this establishes the end point of the titration.

The titration is carried out at a pH of 10, in an $NH_3-NH_4^+$ buffer, which keeps the EDTA (H_4Y) mainly in the half-neutralized form, H_2Y^{2-}, where it complexes the Group IIA ions very well but does not tend to react as readily with other cations such as Fe^{3+} that might be present as impurities in the water. Taking H_4Y and H_3In as the formulas for EDTA and Eriochrome Black T respectively, the equations for the reactions which occur during the titration are:

(main reaction) $H_2Y^{2-}(aq) + Ca^{2+}(aq) \rightarrow CaY^{2-}(aq) + 2 H^+(aq)$ (same for Mg^{2+})

(at end point) $H_2Y^{2-}(aq) + MgIn^-(aq) \rightarrow MgY^{2-}(aq) + HIn^{2-}(aq) + H^+(aq)$
              wine red   sky blue

Since the indicator requires a trace of Mg^{2+} to operate properly, we will add a little magnesium ion to each solution and titrate it as a blank.

EXPERIMENTAL PROCEDURE   WEAR YOUR SAFETY GLASSES WHILE PERFORMING THIS EXPERIMENT

Obtain a 50 cm³ buret, a 250 cm³ volumetric flask, and 25 and 50 cm³ pipets from the stockroom.

Put about a half gram of calcium carbonate in a small 50 cm³ beaker and weigh the beaker and contents on the analytical balance. Using a spatula, transfer about 0.4 g of the carbonate to a 250 cm³ beaker and weigh again, determining the mass of the $CaCO_3$ sample by difference.

Add 25 cm³ of distilled water to the large beaker and then, slowly, about 20 drops of 12 M HCl. Cover the beaker with a watch glass and allow the reaction to proceed until all of the solid carbonate has dissolved. Rinse the walls of the beaker down with distilled water from your wash bottle and heat the solution until it just begins to boil. (Be sure not to be confused by the evolution of CO_2 which occurs with the boiling.) Add 50 cm³ of distilled water to the beaker and carefully transfer the solution, using a stirring rod as a pathway, to the volumetric flask. Rinse the beaker several times with small portions of distilled water and transfer each portion to the flask. All of the Ca^{2+} originally in the beaker should then be in the volumetric flask, and the solution is one of slightly acidic $CaCl_2$. Fill the volumetric flask with distilled water, adding the last few ml a drop at a time with your wash bottle or medicine dropper. When the bottom of the meniscus is just even with the horizontal mark on the flask, stopper the flask and mix the solution thoroughly by inverting the flask at least a dozen times and shaking at intervals over a period of five minutes.

Clean your buret thoroughly and draw about 200 cm³ of the stock EDTA solution from the carboy into a dry 250 cm³ Erlenmeyer flask. Rinse the buret with a little of the solution at least three times. Drain through the stopcock and then fill the buret with the EDTA solution.

Determine a blank by adding 25 cm³ distilled water and 5 cm³ of the pH 10 buffer to a 250 cm³ Erlenmeyer flask. Add two drops of Eriochrome Black T indicator. The solution should turn blue. Add 15 drops 0.03 M MgCl₂, which should contain enough Mg^{2+} to turn the solution wine red. Read the buret to 0.02 cm³ and add EDTA to the solution until the last tinge of purple just disappears. Read the buret again to determine the volume required for the blank. This volume must be subtracted from the total EDTA volume used in each titration. Save the solution as a reference for the end point in all your titrations.

Pipet three 25 cm³ portions of the Ca^{2+} solution in the volumetric flask into clean 250 cm³ Erlenmeyer flasks. To each flask add 5 cm³ of the pH 10 buffer, 2 drops of indicator, and 15 drops of 0.03 M MgCl₂. Titrate the solution in one of the flasks until its color matches that of your reference solution; the end point is a reasonably good one, and you should be able to hit it within a few drops if you are careful. Read the buret. Refill the buret, read it, and titrate the second solution, then the third.

Your instructor will furnish you a sample of water for hardness analysis. Since the concentration of Ca^{2+} is probably lower than that in the standard calcium solution you prepared, pipet 50 cm³ of the water sample for each titration. As before, add 2 drops of indicator, 5 cm³ of pH 10 buffer, and 15 drops of 0.03 M MgCl₂ before titrating. Carry out as many titrations as necessary to obtain two volumes of EDTA that agree within about 3 per cent. If the volume of EDTA required in the first titration is low due to the fact that the water is not very hard, increase the volume of the water sample so that in succeeding titrations, it takes at least 20 cm³ of EDTA to reach the end point.

DATA AND CALCULATIONS: Determination of the Hardness of Water

Mass of beaker
plus CaCO₃ _____ g

Volume Ca²⁺
solution prepared _____ cm³

Mass of beaker
less sample _____ g

Molarity of Ca²⁺ _____ M

Mass of CaCO₃
sample _____ g

Moles Ca²⁺ in each
aliquot titrated _____ mol

Number of moles CaCO₃ in sample
(Formula mass = 100.1) _____ mol

Standardization of EDTA Solution

Determination of blank:

Initial buret
reading _____ cm³

Final buret
reading _____ cm³

Volume of
blank _____ cm³

Titration:	I	II	III
Initial buret reading	_____ cm³	_____ cm³	_____ cm³
Final buret reading	_____ cm³	_____ cm³	_____ cm³
Volume of EDTA	_____ cm³	_____ cm³	_____ cm³
Volume EDTA used to titrate blank	_____ cm³	_____ cm³	_____ cm³
Volume EDTA used to titrate Ca²⁺	_____ cm³	_____ cm³	_____ cm³

Average volume of EDTA required to titrate Ca²⁺ _____ cm³

$$\text{Molarity of EDTA} = \frac{\text{no. moles Ca}^{2+} \text{ in aliquot} \times 1000}{\text{average volume EDTA required (cm}^3)} = \text{_____ } M$$

229

Continued on following page

Determination of Water Hardness

Titration: I   II   III

Volume of water used   _____ cm^3 _____ cm^3 _____ cm^3

Initial buret reading   _____ cm^3 _____ cm^3 _____ cm^3

Final buret reading   _____ cm^3 _____ cm^3 _____ cm^3

Volume of EDTA   _____ cm^3 _____ cm^3 _____ cm^3

Volume of EDTA used
to titrate blank   _____ cm^3 _____ cm^3 _____ cm^3

Volume of EDTA required
to titrate water   _____ cm^3 _____ cm^3 _____ cm^3

Average volume EDTA
per dm^3 of water   _____ cm^3

No. moles EDTA per dm^3 water _____ = No. moles CaCO$_3$ per dm^3 water

No. grams CaCO$_3$ Water hardness
per dm^3 water   _____ g   (1 ppm = 1 mg/dm^3) _____ ppm CaCO$_3$

Unknown No. _____

ADVANCE STUDY ASSIGNMENT: Determination of the Hardness of Water

1.   A 0.5302 g sample of $CaCO_3$ was dissolved in HCl and the resulting solution diluted to 250.0 cm³ in a volumetric flask. A 25.00 cm³ sample of the solution required 27.36 cm³ of an EDTA solution for titration to the Eriochrome Black T end point.

   a. How many moles of $CaCO_3$ were used?

_____mol

   b. What is the concentration of Ca^{2+} in the 250 cm³ of $CaCl_2$ solution?

_____ M

   c. How many moles of Ca^{2+} are contained in a 25.00 cm³ sample?

_____ mol

   d. How many moles of EDTA are contained in the 27.36 cm³ used for titration?

_____ mol

   e. What is the concentration of the EDTA solution?

_____M

2.   If 100 cm³ of a water sample required 17.16 cm³ of EDTA of the concentration found in Problem 1(e), what is the hardness of the water in terms of ppm $CaCO_3$? (1 ppm = 1 mg/dm³)

_____ ppm $CaCO_3$ **231**

EXPERIMENT

30 • Synthesis of Some Coordination Compounds

Some of the most interesting research in inorganic chemistry has involved the preparation and study of the properties of those substances known as coordination compounds. These compounds, sometimes called complexes, are typically salts that contain *complex ions.* A complex ion is an ion that contains a central metal ion to which are bonded small polar molecules or simple ions; the bonding in the complex ion is through coordinate covalent bonds, which ordinarily are relatively weak.

In this experiment we shall be concerned with the synthesis of four coordination compounds:

A. $[Cu(NH_3)_4] SO_4 \cdot H_2O$ B. $[Co(NH_3)_5 H_2O] Cl_3$

C. $[Co(NH_3)_6] Cl_3$ D. $[Co(NH_3)_5 Cl] Cl_2$

The complex ions in these substances are enclosed in brackets to indicate those species which are bonded to the central ion. In this experiment you will prepare the complex ions by making use of reactions in which substituting *ligands,* or coordinating species, replace other ligands on the central ion. The reactions will usually be carried out in water solution, in which the metallic cation will initially be present in the simple hydrated form; addition of a reagent containing a complexing ligand will result in an exchange reaction of the sort

$$Cu(H_2O)_4^{2+}(aq) + 4 NH_3(aq) \rightleftharpoons Cu(NH_3)_4^{2+}(aq) + 4 H_2O \tag{1}$$

In many reactions involving complex ion formation the rate of reaction is very rapid, so that the thermodynamically stable form of the ion is the one produced. Such reactions obey the law of chemical equilibrium and can thus be readily controlled as to direction by a change in the reaction conditions. Reaction 1 proceeds readily to the right in the presence of NH_3 in moderate concentrations. However, by decreasing the NH_3 concentration, for example by the addition of acid to the system, we can easily regenerate the hydrated copper cation. Complex ions that undergo very fast exchange reactions, such as those in Reaction 1, are called *labile.*

Most but by no means all complex ions are labile. Some complex ions, including some to be studied in this experiment, will exchange ligands only slowly. For such species, called *inert* or *nonlabile,* the complex ion produced in a substitution reaction may be the one which is kinetically rather than thermodynamically favored. Alteration of reaction conditions, perhaps by the addition of a catalyst, may change the relative rates of formation of possible complex products and so change the complex ion produced in the reaction. In your experiment, you will find that under one set of conditions the cobalt complex ion in substance B listed previously will be formed, whereas by addition of a catalyst the complex ion in substance C is produced in the presence of the same ligands.

Many complex ions are highly colored, both in solution and in the solid salt. An easy way to determine whether a complex ion is labile is to note whether a color change occurs in a solution containing the ion when a good complexing ligand is added. You may wish to use this procedure in observing relative rates of substitution reactions involving some of the complex ions you prepare.

The experiment as described here will ordinarily take two weeks. If only one week of laboratory time is available, your instructor will tell you which sections of the experiment you should perform and how he wishes to evaluate your product.

EXPERIMENTAL PROCEDURE

A. Preparation of $[Cu(NH_3)_4]SO_4 \cdot H_2O$

$$Cu(H_2O)_4^{2+}(aq) + SO_4^{2-}(aq) + 4\,NH_3(aq) \rightarrow [Cu(NH_3)_4]SO_4 \cdot H_2O(s) + 3\,H_2O$$

Weigh out 7.0 g of $CuSO_4 \cdot 5H_2O$ on a triple-beam or other rough balance. Weigh the solid either on a piece of paper or in a small beaker and not directly on the balance pan. Transfer the solid copper sulfate to a 125 cm³ Erlenmeyer flask and add 15 cm³ of water. Heat the flask to dissolve the solid, then cool to room temperature.

Carry out the remaining steps under the hood. Add 15 M NH_3 solution, a little at a time, swirling the flask to mix the reagents, until the first precipitate has completely dissolved. All the copper should now be present in solution as the complex ion $Cu(NH_3)_4^{2+}$.

The sulfate salt of this complex cation can be precipitated by the addition of a liquid, such as methyl alcohol, in which $Cu(NH_3)_4SO_4 \cdot H_2O$ is insoluble. Add 10 cm³ of methyl alcohol* to the solution; this should result in the formation of a deep blue precipitate of $Cu(NH_3)_4SO_4 \cdot H_2O$. Filter the solution through a Buchner funnel, using suction. Wash the solid in the funnel by adding two 5 cm³ portions of methyl alcohol. Dry the solid by pressing it between two pieces of filter paper. Put the crystals on a piece of filter paper and let them dry further in the air. When they are thoroughly dry, weigh them on the paper.

B. Synthesis of Aquopentamminecobalt(III) Chloride, $[Co(NH_3)_5H_2O]Cl_3$

$$2\,Co(H_2O)_6^{2+}(aq) + 6\,Cl^-(aq) + 10\,NH_3(aq) + H_2O_2(aq)$$
$$\rightarrow 2\,[Co(NH_3)_5H_2O]Cl_3(s) + 10\,H_2O + 2\,OH^-(aq)$$

Obtain a 5 cm³ sample of cobalt(II) chloride solution from your laboratory instructor. This sample contains 1.0 g of $CoCl_2$. Place the cobalt solution in a 150 cm³ beaker and add 0.2 g of ammonium chloride, NH_4Cl, and 11 cm³ of 15 M NH_3 solution. Pour 5 cm³ of 10 per cent hydrogen peroxide, H_2O_2, solution into a 10 cm³ graduated cylinder. Slowly add the H_2O_2 solution to the $CoCl_2$ solution, stirring constantly. When the bubbling has stopped, place the beaker containing the dark red cobalt solution under the hood and gently aerate for an hour using a glass eyedropper and rubber tubing connected to the air jet. Be careful not to turn on the compressed air so vigorously that the solution splatters out of the beaker.

After the compressed air has been stopped, add 15 cm³ of 12 M HCl to the cobalt solution. If a red precipitate does not form in a few minutes, add 5 cm³ more of the HCl solution. Isolate the red precipitate of $[Co(NH_3)_5H_2O]Cl_3$ by filtration through a Buchner funnel, using suction. Continue to draw air through the precipitate by suction for several minutes after the solvent has been removed to partially dry the solid. Press the solid dry between pieces of filter paper and then transfer it to a fresh piece of filter paper. Weigh the crystals after they are dry.

C. Synthesis of Hexamminecobalt(III) Chloride, $[Co(NH_3)_6]Cl_3$

$$2\,Co(H_2O)_6^{2+}(aq) + 6\,Cl^-(aq) + 12\,NH_3(aq) + H_2O_2(aq)$$
$$\rightarrow 2\,[Co(NH_3)_6]Cl_3(s) + 12\,H_2O + 2\,OH^-(aq)$$

---

*Methyl alcohol (not to be confused with ethyl alcohol) is extremely poisonous. Avoid inhalation and contact with the skin.

Obtain a 5 cm³ sample of cobalt (II) chloride solution (containing 1.0 g $CoCl_2$). Place this solution in a 150 cm³ beaker and add 0.2 gram of ammonium chloride, NH_4Cl, 0.1 gram of activated charcoal (Norite) and 5 cm³ of 15 M NH_3. Into a 10 cm³ graduated cylinder draw 5 cm³ of 10 per cent hydrogen peroxide, H_2O_2, solution. Add the H_2O_2 solution very slowly to the $CoCl_2$ solution, stirring constantly. When the bubbling has stopped, gently heat the cobalt solution in the hood for 5 min with a small Bunsen flame. Allow the solution to stand for 30 min. Remove the activated charcoal by passing the solution through filter paper on a Buchner funnel, using suction. Pour the filtrate into a 150 cm³ beaker and add 20 cm³ of 12 M HCl. Stir the solution thoroughly and isolate the orange precipitate of $[Co(NH_3)_6]Cl_3$ by filtration through the Buchner funnel with suction. Continue to draw air through the funnel for several minutes to help dry the solid. Dry the product further by pressing the crystals between two pieces of filter paper. Let the crystals finish drying on a piece of filter paper and then weigh them.

D. Synthesis of Chloropentamminecobalt(III) Chloride, $[Co(NH_3)_5Cl]Cl_2$

$$[Co(NH_3)_5H_2O]Cl_3(s) \rightarrow [Co(NH_3)_5Cl]Cl_2(s) + H_2O(g)$$

Place the aquopentamminecobalt(III) chloride which was prepared in Part B on a small watch glass. Place the watch glass on a piece of asbestos-covered wire gauze supported on a ring stand. Adjust the height of the ring so that the gauze is about 5 cm above the top of a Bunsen burner tip. Spread the $[Co(NH_3)_5H_2O]Cl_3$ into a thin layer on the watch glass, using a spatula. Ignite the Bunsen burner and adjust the flame until the blue flame is about 2 cm high. Heat the wire gauze in the center (gently!) and observe the $[Co(NH_3)_5H_2O]Cl_3$. When the center of the red solid begins to darken, remove the flame. Move any remaining red solid to the center of the watch glass with a spatula, spreading the solid thin! Gently reheat the watch glass until only the purple product, $[Co(NH_3)_5Cl]Cl_2$, is left. Weigh the final product.

E. Relative Lability of Complex Ions

Dissolve a small portion (about 0.5 g) of the complexes from Parts A and C, $[Cu(NH_3)_4]SO_4 \cdot H_2O$ and $[Co(NH_3)_6]Cl_3$, in a little water. Note the color and observe the effect of the addition of a few drops of 12 M HCl.

CAUTION: In these experiments you will be using 15 M NH_3, 12 M HCl, and a 10% solution of H_2O_2. Avoid getting these reagents on your skin or clothing and don't breathe the vapors of NH_3 or HCl. Wash off any spilled reagents thoroughly with water.

DATA AND OBSERVATIONS: Synthesis of Some Coordination Compounds

	A	B	C	D
Yield of products	_____ g	_____ g	_____ g	_____ g
Theoretical yield of products	_____ g	_____ g	_____ g	_____ g
Percentage yield of products	_____ %	_____ %	_____ %	_____ %

Observations Regarding Lability of Complexes (Part E)

	color of solid	color in H_2O solution	color on addition of HCl
$Cu(NH_3)_4^{2+}$	_____	_____	_____
$Co(NH_3)_6^{3+}$	_____	_____	_____

Conclusions

What evidence, if any, do you have that any of the formulas we have given for the compounds prepared in Parts A, B, C, and D are correct, either as to the nature of the atoms or groups present or as to the number of the groups present?

ADVANCE STUDY ASSIGNMENT: **Synthesis of Some Coordination Compounds**

1. a. Calculate the theoretical yields of all compounds to be prepared in the first four parts of this experiment. The metal ion in all cases is the limiting reagent.

A.

_____ g

B.

_____ g

C.

_____ g

D.

_____ g

   b. How would you establish that the metal ion is the limiting reagent?

2. How would you name the compound $[Cu(NH_3)_4]SO_4 \cdot H_2O$ prepared in Part A?

3. What species would you expect to be present in aqueous solutions of each of the complexes to be prepared in these experiments?

A.   **C.**

B.   **D.**

ELECTROCHEMISTRY

EXPERIMENT

31 • Determination of a Gram Equivalent Mass by Electrolysis

The gram equivalent mass of an element was defined in earlier experiments as that mass of the element which will combine with 8.000 g of oxygen or with one gram equivalent mass of any other element. Experiments 4, 5, and 6 dealt with the determination of gram equivalent masses by straightforward chemical means.

Experimentally we find that the gram equivalent mass of an element can also be related in a fundamental way to the chemical effects observed in that phenomenon known as *electrolysis*. As you know, some liquids, because they contain ions, will conduct an electric current. If the two terminals on a storage battery, or any other source of D.C. voltage, are connected through metal electrodes to a conducting liquid, an electric current will pass through the liquid and chemical reactions will occur at the two metal electrodes; in this experiment electrolysis is said to occur, and the liquid is said to be electrolyzed.

At the electrode connected to the *negative* pole of the battery, a *reduction* reaction will invariably be observed. In this reaction electrons will usually be accepted by one of the species present in the liquid, which, in the experiment we shall be doing, will be an aqueous solution. The species reduced will ordinarily be a metallic cation or the H^+ ion or possibly water itself; the reaction which is actually observed will be the one that occurs with the least expenditure of electrical energy, and will depend on the composition of the solution. In the electrolysis cell we shall study, the reduction reaction of interest will occur in an acid medium; hydrogen gas will be produced by the reduction of hydrogen ion:

$$2\,H^+(aq) + 2\,e^- \rightarrow H_2(g) \tag{1}$$

In this reduction reaction, which will occur at the negative pole, or *cathode*, of the cell, for every H^+ ion reduced *one* electron will be required, and for every molecule of H_2 formed, *two* electrons will be needed.

Ordinarily in chemistry we deal not with individual ions or molecules but rather with moles of substances. In terms of moles, we can say that, by Equation 1,

The reduction of one mole of H^+ ion requires one mole of electrons

The production of one mole of $H_2(g)$ requires two moles of electrons

A mole of electrons is a fundamental amount of electricity in the same way that a mole of pure substance is a fundamental unit of matter, at least from a chemical point of view. The amount of a species which will react with a *mole* of *electrons* is equal to the *gram equivalent mass* of that species. Since one mole of electrons will reduce one mole of H^+ ion, we say that the gram equivalent mass of hydrogen is 1.008 g, the mass of one mole of H^+ ion (or one-half mole of $H_2(g)$). To form one mole of $H_2(g)$ one would have to pass two moles of electrons through the electrolysis cell.

In the electrolysis experiment we will perform we will measure the volume of hydrogen gas produced under known conditions of temperature and pressure. By using the ideal gas law we will be able to calculate how many moles of H_2 were formed, and hence how many moles of electrons of electricity passed through the cell.

At the positive pole of an electrolysis cell (the metal electrode that is connected to the + terminal of the battery), an *oxidation* reaction will occur, in which some species will give up electrons. This reaction, which takes place at the *anode* in the cell, may involve again an ionic or neutral species in the solution or the metallic electrode itself. In the cell that you will be studying, the pertinent oxidation reaction will be that in which a metal under study will participate:

$$M(s) \rightarrow M^{n+}(aq) + ne^- \tag{2}$$

During the course of the electrolysis the atoms in the metal electrode will be converted to metallic cations and will go into the solution. The mass of the metal electrode will decrease, depending on the amount of electricity passing through the cell and the nature of the metal. In order to oxidize one mole, or one gram atomic mass, of the metal, it would take n moles of electrons, where n is the charge on the cation which is formed. By definition, one mole of electrons of electricity would cause one gram equivalent mass, GEM, of metal to go into solution. The gram atomic mass, GAM, and the gram equivalent mass of the metal are clearly related by the equation:

$$GAM = GEM \times n \tag{3}$$

In an electrolysis experiment, since n is not determined independently, it is not possible to find the gram atomic mass of a substance. It is possible, however, to find gram equivalent masses of many metals, and that will be our purpose.

The general method we will use is implied by the discussion. We will oxidize a sample of an unknown metal at the positive pole of an electrolysis cell, weighing the metal before and after the electrolysis and so determining its loss in mass. We will use the same amount of electricity, the same number of electrons, to reduce hydrogen ion at the negative pole of an electrolysis cell. From the volume of H_2 gas which is produced under known conditions we can calculate the number of moles of H_2 formed, and hence the number of moles of electrons which passed through the cell. (See Problem 2 in the advance study assignment for illustrative data.) The gram equivalent mass of the metal is then calculated as the amount of metal which would be oxidized if one mole of electrons were used.

EXPERIMENTAL PROCEDURE   WEAR YOUR SAFETY GLASSES WHILE PERFORMING THIS EXPERIMENT

Obtain from the stockroom a buret and a sample of a metal unknown. Weigh the metal sample on the analytical balance.

Set up the electrolysis apparatus as indicated in Figure 31.1. There should be about $100 \ cm^3 \ 1 \ M \ H_2SO_4$ in the beaker with the gas buret. Immerse the end of the buret in the acid and attach a length of rubber tubing to its upper end. Open the stopcock on the buret and, with suction, carefully draw the acid up to the top of the graduations. Close the stopcock. Insert the bare coiled end of the heavy copper wire up into the end of the buret; all but the coil end of the wire should be covered with watertight insulation. Check the acid level after a few minutes to make sure the stopcock does not leak.

The hydrogen cell is connected to the other cell by a piece of heavy nichrome wire, which serves as an anode in one cell and a cathode in the other. The reactions that occur at the nichrome electrodes are not of concern in this experiment. The second 125 cm³ beaker should contain about 100 cm³ 0.5 M KNO_3. The anode in that beaker is made from your sample of unknown metal. Connect the metal to the + pole of the power source with an alligator clip and immerse the metal but not the clip in the KNO_3 solution. Read and record the liquid level in the buret.

Begin electrolysis by connecting the copper electrode to the negative pole of the power source. Hydrogen gas should immediately begin to bubble from the copper cathode. Collect the gas until about 50 cm³ have been produced. At that point, stop the electrolysis by disconnecting the cell from the power source. Record the liquid level in

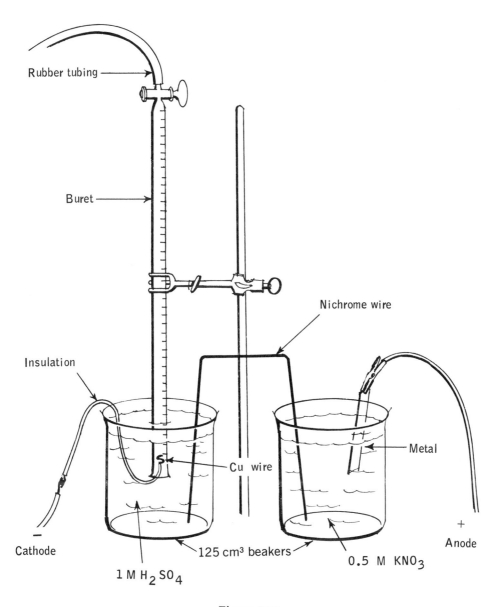

Figure 31.1

the buret. (In some cases a cloudiness will appear in the KNO_3 solution during the electrolysis; this is caused by formation of a metal hydroxide, and will have no adverse effect on the experiment.) Open the stopcock on the buret and again draw the acid up to the top of the graduations. Read the buret, reconnect the electrical wires, and generate another 50 cm³ of hydrogen as before.

Record the liquid level in the buret and the temperature and barometric pressure in the laboratory. Remove the metal anode from the electrolysis cell and rinse it with 0.1 M or, if necessary, 1 M acetic acid; if it has a flaky coating, scrape this off and rinse the electrode again; dry it by immersing it in acetone and letting the acetone evaporate in the air. Weigh the metal anode.

DATA AND CALCULATIONS: **Determination of a Gram Equivalent Mass by Electrolysis**

Mass of metal anode   _____ g

Mass of anode after electrolysis   _____ g

Initial buret reading   _____ cm³

Buret reading after first electrolysis   _____ cm³

Buret reading after refilling   _____ cm³

Buret reading after second electrolysis   _____ cm³

Barometric pressure (1 mm Hg = 0.1333 kPa)   _____ kPa

Temperature t _____ °C

Vapor pressure of H_2O at t _____ kPa

Total volume of H_2 produced, V _____ cm³

Temperature T _____ K

Pressure exerted by dry H_2: $P = P_{Bar} - VP_{H_2O}$
(ignore any pressure effect due to liquid levels in buret)   _____ kPa

No. moles H_2 produced, n
(use Ideal Gas Law, $PV = nRT$)   _____ mol

No. of moles of electrons passed   _____

Loss in mass by anode   _____ g

Gram equivalent mass of metal   _____ g

Unknown metal number   _____

ADVANCE STUDY ASSIGNMENT: Determination of a Gram Equivalent
                             Mass by Electrolysis

1.   In the electrolysis experiment, two cells are used. Why is it possible to relate the
effect at the cathode of one cell to the anode of the other?

2.   A current passing through an electrolytic cell such as the one employed in this
experiment liberates 85.30 cm³ of H_2 at the cathode and causes a mass loss of 0.2124 g at
the metal anode. If the barometric pressure is 99.0 kPa, and the temperature in the
laboratory is 24°C, what is the gram equivalent mass of the metal? At 24°C the vapor
pressure of water is 3.0 kPa.

P_{H_2} = _____ kPa

V_{H_2} = _____ cm³ = _____ dm³

T = _____ K

n_{H_2} = _____ mol

1 mol H_2 requires _____ mol electrons

No. moles e⁻ passed = _____ = No. GEM metal electrolyzed

No. g metal oxidized = _____ g

GEM metal = $\dfrac{\text{No. g}}{\text{No. moles e}^-}$ = _____ = _____ g

3.   In this experiment we make the assumption that the reactions occurring at the
cathode and anode are the only reactions which take place. That is, we assume that the
electrodes are 100% efficient. If other reactions also occur at the electrodes of interest,
the results of the experiment will be incorrect. What effect would the evolution of O_2
occurring simultaneously with the oxidation of the metal anode have on the value of the
GEM of the metal which would be obtained?

EXPERIMENT

32 • Voltaic Cell Measurements

Many chemical reactions can be classified as oxidation-reduction reactions, since they involve the oxidation of one species and the reduction of another. Such reactions can conveniently be considered as the result of two half reactions, one of oxidation and the other reduction. In the case of the oxidation-reduction reaction

$$Zn(s) + Pb^{2+}(aq) \rightarrow Zn^{2+}(aq) + Pb(s)$$

which would occur if a piece of metallic zinc were put into a solution of lead nitrate, the two reactions would be

$$Zn(s) \rightarrow Zn^{2+}(aq) + 2e^- \quad \text{oxidation}$$

$$2e^- + Pb^{2+}(aq) \rightarrow Pb(s) \quad \text{reduction}$$

The tendency for an oxidation-reduction reaction to occur can be measured if the two reactions are made to occur in separate regions connected by a barrier that is porous to ion movement. An apparatus called a *voltaic cell* in which this reaction might be carried out under this condition is shown in Figure 32.1.

If we connect a voltmeter between the two electrodes we will find that there is a voltage between them. The magnitude of the voltage is a direct measure of the driving

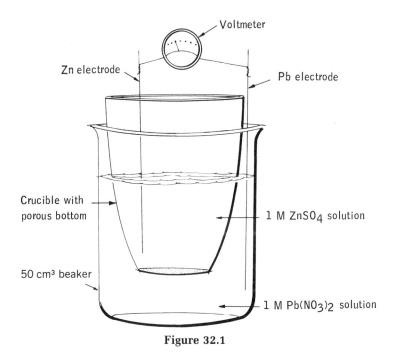

Figure 32.1

force or thermodynamic tendency of the oxidation-reduction reaction to occur.

If we study several oxidation-reduction reactions we find that the voltage of each associated voltaic cell can be considered to be the sum of a voltage for the oxidation reaction and a voltage for the reduction reaction. In the Zn, $Zn^{2+} \parallel Pb^{2+}$, Pb cell we have been discussing, for example,

$$E_{cell} = E_{Zn, Zn^{2+} \text{ oxidation reaction}} + E_{Pb^{2+}, \text{ Pb reduction reaction}} \tag{1}$$

By convention, the negative electrode in a voltaic cell is taken to be the one from which electrons are emitted (i.e., where oxidation occurs).

Since any cell voltage is the sum of two electrode voltages, it is not possible, by measuring cell voltages, to determine individual absolute electrode voltages. However, if a value of voltage is arbitrarily assigned to one electrode reaction, then other electrode voltages can be given definite values, based on the assigned value. The usual procedure is to assign a value of 0.0000 volts to the standard voltage for the electrode reaction

$$2\,H^+(aq) + 2\,e^- \rightarrow H_2(g); \quad E_{H^+, H_2 \text{ red}} = 0.0000 \text{ V}$$

For the Zn, $Zn^{2+} \parallel H^+$, H_2 cell, the measured voltage is 0.76 V, and the zinc electrode is negative. Zinc metal is therefore oxidized, and the cell reaction must be

$$Zn(s) + 2\,H^+(aq) \rightarrow Zn^{2+}(aq) + H_2(g); \quad E_{cell} = 0.76 \text{ V}$$

Given this information, one can readily find the voltage for the oxidation of Zn to Zn^{2+}.

$$E_{cell} = E_{Zn, Zn^{2+} \text{ oxid}} + E_{H^+, H_2 \text{ red}}$$

$$0.76 \text{ V} = E_{Zn, Zn^{2+} \text{ oxid}} + 0.00 \text{ V}; \quad E_{Zn, Zn^{2+} \text{ oxid}} = +0.76 \text{ V}$$

If the voltage for a half reaction is known, the voltage for the reverse reaction can be obtained by changing the sign. For example:

$$\textit{if } E_{Zn, Zn^{2+} \text{ oxid}} = +X \text{ volts}, \quad \textit{then } E_{Zn^{2+}, Zn \text{ red}} = -X \text{ volts}$$

$$\textit{if } E_{Pb^{2+}, Pb \text{ red}} = +Y \text{ volts}, \quad \textit{then } E_{Pb, Pb^{2+} \text{ oxid}} = -Y \text{ volts}$$

In the first part of this experiment you will measure the voltages of several different cells. By arbitrarily assigning the voltage of a particular half reaction to be 0.00 V, you will then be able to calculate the voltages corresponding to all of the various half reactions that occurred in your cells.

In our discussion so far we have not considered the possible effects of such system variables as temperature, voltage at the liquid-liquid junction, size of metal electrodes, and concentrations of solute species. Although temperature and liquid junctions do have a definite effect on cell voltages, taking account of their influence involves rather complex thermodynamic concepts and is usually not of concern in any elementary course. The size of a metal electrode has no appreciable effect on electrode voltage, although it does relate directly to the capacity of the cell to produce useful electrical energy. In this experiment we will operate the cells so that they deliver essentially no energy but exert their maximum voltages.

The effect of solute ion concentrations is important and can be described relatively easily. For the cell reaction at 25°C:

$$aA(s) + bB^+(aq) \rightarrow cC(s) + dD^{2+}(aq)$$

$$E_{cell} = E_{cell}^\circ - \frac{0.06}{n} \log \frac{(\text{conc. } D^{2+})^d}{(\text{conc. } B^+)^b} \tag{2}$$

where $E°_{cell}$ is a constant for a given reaction and is called the standard cell voltage, and n is the number of electrons in either electrode reaction. (Strictly speaking, in Equation (2) the so-called activities of the ions should be used; see Experiment 19 for a further discussion of the use of activities in dealing with equilibrium systems.)

By Equation 2 you can see that the measured cell voltage, E_{cell}, will equal the standard cell voltage if the molarities of D^{2+} and B^+ are both unity, or, if d equals b, if they are simply equal to each other. We will carry out our experiments under such conditions that the cell voltages you observe will be very close to the standard voltages given in the tables in your chemistry text.

Considering the $Cu,Cu^{2+} \parallel Ag^+,Ag$ cell as a specific example, the observed cell reaction would be

$$Cu(s) + 2\,Ag^+(aq) \rightarrow Cu^{2+}(aq) + 2\,Ag(s)$$

For this cell, Equation 2 takes the form

$$E_{cell} = E°_{cell} - \frac{0.06}{2} \log \frac{conc.\ Cu^{2+}}{(conc.\ Ag^+)^2} \tag{3}$$

In the equation n is 2 because in the cell reaction, two electrons are transferred in each of the two half reactions. $E°$ would be the cell voltage when the copper and silver salt solutions are both $1\ M$, since then the logarithm term is equal to zero.

If we decrease the Cu^{2+} concentration, keeping that of Ag^+ at $1\ M$, the voltage of the cell will go up by about 0.03 volts for every factor of ten by which we decrease conc. Cu^{2+}. Ordinarily it is not convenient to change concentrations of an ion by several orders of magnitude, so in general, concentration effects in cells are relatively small. However, if we should add a complexing or precipitating species to the copper salt solution, the value of conc. Cu^{2+} would drop drastically, and the voltage change would be appreciable. In the experiment we will illustrate this effect by using NH_3 to complex the Cu^{2+}. Using Equation 3, we can actually calculate conc. Cu^{2+} in the solution of its complex ion.

In an analogous experiment we will determine the solubility product of AgCl. In this case we will surround the Ag electrode in a $Cu,Cu^{2+} \parallel Ag^+,Ag$ cell with a solution of known Cl^- ion concentration which is saturated with AgCl. From the measured cell voltage, we can use Equation 3 to calculate the very small value of conc. Ag^+ in the chloride-containing solution. Knowing the concentrations of Ag^+ and Cl^- in a solution in equilibrium with AgCl(s) allows us to find K_{sp} for AgCl.

EXPERIMENTAL PROCEDURE

WEAR YOUR SAFETY GLASSES WHILE PERFORMING THIS EXPERIMENT

You may work in pairs in this experiment.

A. Cell Voltages. In this experiment you will be working with these seven electrode systems:

$Ag^+, Ag(s)$	$Br_2(l), Br^-, Pt$
$Cu^{2+}, Cu(s)$	$Cl_2(g, 1\ atm), Cl^-, Pt$
Fe^{3+}, Fe^{2+}, Pt	$I_2(s), I^-, Pt$
$Zn^{2+}, Zn(s)$	

Your purpose will be to measure enough cell voltages to allow you to determine the electrode voltages of each electrode by comparing it with an arbitrarily chosen electrode voltage.

Using the apparatus shown in Figure 32.1, set up a voltaic cell involving any two of the electrodes in the list. The solute ion concentrations may be assumed to be one molar and all other species may be assumed to be at unit activity, so that the voltages of the cells you set up will be essentially the standard voltages. Measure the cell voltage and record it along with which electrode has negative polarity.

In a similar manner set up and measure the cell voltages and polarities of other cells, sufficient in number to include all of the electrode systems on the list at least once. Do not combine the silver electrode system with any of the halogen electrode systems, since a precipitate will form; any other combinations may be used. The data from this part of the experiment should be entered in the first three columns of the table in Part A.1 of your report.

B.   Effect of Concentration on Cell Voltages

1. COMPLEX ION FORMATION.   Set up the Cu, Cu^{2+} ‖ Ag^+, Ag cell, using 10 cm³ of the $CuSO_4$ solution in the crucible and the solution of $AgNO_3$ in the beaker. Measure the voltage of the cell. While the voltage is being measured, add 10 cm³ of 6 M NH_3 to the $CuSO_4$ solution, stirring carefully with your stirring rod. Measure the voltage when it becomes steady.

2. DETERMINATION OF THE SOLUBILITY PRODUCT OF AgCl.   Remove the crucible from the cell you have just studied and discard the solution of $Cu(NH_3)_4^{2+}$. Clean the crucible by drawing a little 6 M NH_3 through it, using the adapter and suction flask. Then draw some distilled water through it. Reassemble the Cu-Ag cell, this time using the beaker for the Cu-$CuSO_4$ electrode system. Immerse the Ag electrode in the crucible in 1 M KCl; add a drop of $AgNO_3$ solution to form a little AgCl, so that an equilibrium between Ag^+ and Cl^- can be established. Measure the voltage of this cell, noting which electrode is negative. In this case conc. Ag^+ will be very low, which will decrease the voltage of the cell to such an extent that its polarity may change from that observed previously.

DATA AND CALCULATIONS: Voltaic Cell Measurements

A. 1. Cell Voltages

Electrode systems used in Cell	Cell voltage, E°_{cell} (volts)	Negative electrode	Oxidation reaction	$E^\circ_{oxidation}$ in volts	Reduction reaction	$E^\circ_{reduction}$ in volts
1.						
2.						
3.						
4.						
5.						
6.						
7.						

CALCULATIONS

A. Noting that oxidation occurs at the negative pole in a cell, write the oxidation reaction in each of the cells. The other electrode system must undergo reduction; write the reduction reaction which occurs in each cell.

B. Assume that $E^\circ_{Ag^+,\ Ag} = 0.00$ volts (whether in reduction or oxidation). Enter that value in the table for all of the silver electrode systems you used in your cells. Since $E^\circ_{cell} = E^\circ_{oxidation} + E^\circ_{reduction}$, you can calculate E° values for all the electrode systems in which the Ag, Ag^+ system was involved. Enter those values in the table.

C. Using the values and relations in B and taking advantage of the fact that for any given electrode system, $E^\circ_{oxidation} = -E^\circ_{reduction}$, complete the table of E° values. The best way to do this is to use one of the E° values you found in B in another cell with that electrode system. That voltage, along with E°_{cell}, will allow you to find the voltage of the other electrode. Continue this process with other cells until all the electrode voltages have been determined.

Continued on following page

253

A.2. Table of Electrode Voltages

In Table A.1, you should have a value for $E°_{red}$ or $E°_{oxid}$ for each of the electrode systems you have studied. Remembering that for any electrode system, $E°_{red} = -E°_{oxid}$, you can find the value for $E°_{red}$ for each system. List those voltages in the left column of the table below in order of decreasing value.

$E°_{reduction}$ ($E°_{Ag^+, Ag} = 0.00$ volts)	Electrode reaction in reduction	$E°_{reduction}$ ($E°_{H^+, H_2} = 0.00$ volts)
_____	_____	_____
_____	_____	_____
_____	_____	_____
_____	_____	_____
_____	_____	_____
_____	_____	_____
_____	_____	_____

The electrode voltages you have determined are based on $E°_{Ag^+, Ag} = 0.00$ V. The usual assumption is that $E°_{H^+, H_2} = 0.00$ V, under which conditions $E°_{Ag^+, Ag\ red} = 0.80$ V. Convert from one base to the other by adding 0.80 V to each of the electrode voltages and enter these values in the third column of the table.

Why are the values of $E°_{red}$ on the two bases related to each other in such a simple way?

B. Effect of Concentration on Cell Voltages

1. Complex ion formation:

Voltage, $E°_{cell}$, before addition of 6 M NH_3 _____ V

Voltage, E_{cell}, after $Cu(NH_3)_4^{2+}$ formed _____ V

Continued on following page

Given Equation 3

$$E_{cell} = E°_{cell} - \frac{0.06}{2} \log \frac{conc.\ Cu^{2+}}{(conc.\ Ag^+)^2} \tag{3}$$

calculate the residual concentration of free Cu^{2+} ion in equilibrium with $Cu(NH_3)_4^{2+}$ in the solution in the crucible.

conc. $Cu^{2+} =$ _____ M

2. Solubility product of AgCl:

Voltage, $E°_{cell}$, of the Cu, Cu^{2+} ‖ Ag^+, Ag cell (from B.1)

_____V Negative electrode_____

Voltage, E_{cell}, with 1 M KCl present   _____V Negative electrode _____

Using Equation 3, calculate (conc. Ag^+) in the cell, where it is in equilibrium with 1 M Cl^- ion. (E_{cell} in Equation 3 is the *negative* of the measured value if the polarity is not the same as in the standard cell.)

conc. $Ag^+ =$ _____ M

Determine K_{sp} AgCl on the basis of your results.

$K_{sp} =$ _____

ADVANCE STUDY ASSIGNMENT: Voltaic Cell Measurements

1. Write the electrode reactions associated with the following oxidation-reduction reaction: $Cd(s) + 2\ Ag^+(aq) \rightarrow 2\ Ag(s) + Cd^{2+}(aq)$

2. Describe how you would make a voltaic cell to study this reaction.

3. If the reaction as written occurred spontaneously in the cell, which electrode would be negative?

4. How would the cell voltage change if the reaction above were written for two moles of $Cd(s)$?

5. If the cell voltage measured is 1.20 V and the reaction is that described in Problem 1, what would be the *reduction* voltage associated with the Cd^{2+},Cd electrode if

    a. the Ag,Ag^+ electrode is assigned a voltage of 0.000 V?
    b. the Ag,Ag^+ electrode is assigned a reduction voltage of 0.80 V?

                                                    a. _____ V

                                                    b. _____ V

6. Referring to Equation 3, what voltage would you expect to obtain for a Cu, Cu^{2+} ‖ Ag^+, Ag cell in which the solution surrounding the Ag electrode is $1\ M$ in I^- and saturated with AgI (K_{sp} AgI $= 1 \times 10^{-16}$)? Assume $E°_{cell} = 0.46$ V.

                                                   _____ V

EXPERIMENT

33 • Preparation of Copper(I) Chloride

Oxidation-reduction reactions are, like precipitation reactions, often used in the preparation of inorganic substances. In this experiment we will employ a similar series of reactions to prepare one of the less commonly encountered salts of copper, copper(I) chloride. Most copper compounds contain copper(II), but copper(I) is present in a few slightly soluble or complex copper salts.

The process of synthesis of CuCl we will use begins by dissolving copper metal in nitric acid:

$$Cu(s) + 4\,H^+(aq) + 2\,NO_3^-(aq) \rightarrow Cu^{2+}(aq) + 2\,NO_2(g) + 2\,H_2O \tag{1}$$

The solution obtained is treated with sodium carbonate in excess, which neutralizes the remaining acid with evolution of CO_2 and precipitates Cu(II) as the carbonate:

$$2\,H^+(aq) + CO_3^{2-}(aq) \rightleftharpoons (H_2CO_3)\,(aq) \rightleftharpoons CO_2(g) + H_2O \tag{2}$$

$$Cu^{2+}(aq) + CO_3^{2-}(aq) \rightleftharpoons CuCO_3(s) \tag{3}$$

The $CuCO_3$ will be purified by filtration and washing and dissolved in hydrochloric acid. Copper metal added to the highly acidic solution then reduces the Cu(II) to Cu(I) and is itself oxidized to Cu(I). In the presence of excess chloride, the copper will be present as a $CuCl_4^{3-}$ complex ion. Addition of this solution to water destroys the complex, and white CuCl precipitates.

$$CuCO_3(s) + 2\,H^+(aq) + 4\,Cl^-(aq) \rightarrow CuCl_4^{2-}(aq) + CO_2(g) + H_2O \tag{4}$$

$$CuCl_4^{2-}(aq) + Cu(s) + 4\,Cl^-(aq) \rightarrow 2\,CuCl_4^{3-}(aq) \tag{5}$$

$$CuCl_4^{3-}(aq) \xrightarrow{H_2O} CuCl(s) + 3\,Cl^-(aq) \tag{6}$$

Since CuCl is readily oxidized, due care must be taken to minimize its exposure to air during its preparation and while it is being dried.

EXPERIMENTAL PROCEDURE

WEAR YOUR SAFETY GLASSES WHILE PERFORMING THIS EXPERIMENT

Obtain a 1 g sample of copper metal, a Buchner funnel, and a filter flask from the stockroom. Weigh the copper metal on the top loading or triple beam balance to 0.1 g.

Put the metal in a 150 cm³ beaker and *under a hood* add 5 cm³ 15 M HNO₃. Brown NO₂ gas will be evolved and an acid blue solution of Cu(NO₃)₂ produced. If it is necessary, you may warm the beaker gently with a Bunsen burner to dissolve all of the copper. When all of the copper is in solution, add 50 cm³ of water to the solution and allow it to cool.

Weigh out about 5 g of sodium carbonate in a small beaker on a rough balance. Add small amounts of the Na₂CO₃ to the solution with your spatula, adding the solid as necessary when the evolution of CO₂ subsides. Stir the solution to expose it to the solid.

259

When the acid is neutralized, a blue-green precipitate of $CuCO_3$ will begin to form. At that point, add the rest of the Na_2CO_3, stirring the mixture well to ensure complete precipitation of the copper carbonate.

Transfer the precipitate to the Buchner funnel and use suction to remove the excess liquid. Use your rubber policeman and a spray from your wash bottle to make a complete transfer of the solid. Wash the precipitate well with distilled water with suction on, then let it remain on the filter paper with suction on for a minute or two.

Remove the filter paper from the funnel and transfer the solid $CuCO_3$ to the 150 cm³ beaker. Add 25 cm³ water and then 10 cm³ 12 M HCl slowly to the solid, stirring continuously. When the $CuCO_3$ has all dissolved, add 1.5 g Cu foil cut in small pieces to the beaker and cover it with a watch glass.

Heat the mixture in the beaker to the boiling point and keep it at that temperature, just simmering, for 30 to 40 min. It may be that the dark-colored solution which forms will clear to a yellow color before that time is up, and if it does, you may stop heating and proceed with the next step.

While the mixture is heating, put 150 cm³ distilled water in a 400 cm³ beaker and put the beaker in an ice bath. Cover the beaker with a watch glass. After you have heated the acidic Cu-$CuCl_2$ mixture for 40 min or as soon as it turns light-colored, carefully decant the hot liquid into the beaker of water, taking care not to transfer any of the excess Cu metal to the beaker. White crystals of CuCl should form. Continue to cool the beaker in the ice bath to promote crystallization and to increase the yield of solid.

Pour 25 cm³ glacial acetic acid *(careful, it's a caustic reagent)* and 25 cm³ acetone into separate small beakers. Filter the crystals of CuCl in the Buchner funnel using suction. Swirl the beaker to aid in transferring the solid to the funnel. Just as the last of the liquid is being pulled through, wash the CuCl with one third of the glacial acetic acid. Rinse the last of the CuCl into the funnel with another portion of the acid and use the final third to rewash the solid. Turn off suction and add one half of the acetone to the funnel; wait about ten seconds and turn on the suction. Repeat this operation with the other half of the acetone. Draw air through the sample for a few minutes to dry it. If you have properly washed the solid, it will be pure white; if the moist compound is allowed to come into contact with air, it will tend to turn pale green, due to oxidation of Cu(I) to Cu(II). Weigh the CuCl in a previously weighed beaker to 0.1 g. Show your sample to your instructor for his evaluation.

DATA AND RESULTS: Preparation of CuCl

Mass of Cu sample   _____ g

Mass of watch glass   _____ g

Mass of beaker plus CuCl   _____ g

Mass of CuCl prepared   _____ g

Theoretical yield   _____ g

Percentage yield   _____ %

ADVANCED STUDY ASSIGNMENT: Preparation of Copper(I) Chloride

1.   The Cu^{2+} ions in this experiment are produced from the reaction of 1.0 g of copper foil with excess nitric acid. How many moles of Cu^{2+} are produced?

_____mol Cu^{2+}

2.   Why isn't hydrochloric acid used in a direct reaction with copper wire to prepare the $CuCl_2$ solution?

3.   How many grams of metallic copper are required to react with the number of moles of Cu^{2+} calculated in (1) to form the CuCl?

_____ g Cu

4.   What is the maximum mass of CuCl that can be prepared from the reaction sequence of this experiment using 1.0 g of Cu foil to prepare the Cu^{2+} solution?

_____ g CuCl

SECTION FOURTEEN
NUCLEAR CHEMISTRY

EXPERIMENT

34 • Determination of the Half-life of a Radioactive Isotope

Some atomic nuclei are radioactive; that is to say, they can spontaneously undergo reactions in which they converted to different nuclei, which may or may not belong to different elements. For example $^{20}_{9}F$ nuclei may spontaneously emit electrons in a reaction in which $^{20}_{10}Ne$ nuclei are produced:

$$^{20}_{9}F \rightarrow {}^{20}_{10}Ne + {}^{0}_{-1}e$$

Reactions such as this differ from ordinary chemical reactions in several ways. Typically they produce electrons, He^{2+} nuclei, γ-rays, or some combination of these, all having very high associated energies. The reactions ordinarily involve relatively few atoms, and we usually study them by measuring the high energy particles or radiations that are emitted. Neither the temperature nor the state of chemical combination of the substances containing the active nuclei affect appreciably either the rate at which the reaction occurs or the energies of the emitted species.

The rate at which given radioactive nuclei decay depends only on the kind and number of nuclei present. In any fixed time interval Δt a certain fraction of the active nuclei will undergo reaction according to the equation

$$\frac{\Delta n}{n} = -k\Delta t \qquad (1)$$

where n is the number of active nuclei in the sample and Δn is the change in the number of active nuclei (or minus the number undergoing decay) in the time interval Δt. Equation 1 can be rewritten to read

$$\frac{\Delta n}{\Delta t} = -kn \qquad (2)$$

which tells us that the rate of radioactive decay, $\Delta n/\Delta t$, is proportional to the number of active nuclei present. The proportionality constant k is called the rate constant for the decay, and the minus sign indicates that the number of active nuclei in the sample decreases with time.

265

By the methods of the calculus, Equation 2 can be integrated to produce an equation that allows us to calculate the number of active nuclei in a sample at any given time, given the initial number of nuclei n_0 and the associated decay constant:

$$\log_{10} n = \log_{10} n_0 - \frac{kt}{2.3} \qquad (3)$$

The time t required for half of the active nuclei in a sample to decay is called the half-life of the nuclei, $t_{1/2}$. If, in Equation 3, $t = t_{1/2}$, then n equals $n_0/2$, and it becomes clear that $t_{1/2}$ and k are related by the equation

$$\log_{10} 1/2 = \frac{-kt_{1/2}}{2.3} \qquad \text{or} \qquad t_{1/2} = \frac{-0.301 \times 2.303}{-k} = \frac{0.693}{k} \qquad (4)$$

If by some means we can find k for a given active isotope, we can find its half-life by Equation 4.

Devices have been developed that are sensitive to the high-energy particles emitted in the nuclear reactions that occur during radioactive decay processes. Upon entering such a device, usually called a counter, a high-energy particle causes an electric discharge in the counter. This discharge is automatically recorded on a digital readout meter. In a typical experiment the active sample is placed near the counter, and the number of particles emitted in the direction of the detector is determined. If the number of particles entering the counter is measured as a function of time, we can measure in a relative way the number of active atoms in the whole sample; for example, if after 10 min the number of counts per minute is only 80 per cent of the number at the beginning of the experiment, there are only 80 per cent as many active nuclei in the sample as there were at the start. This means that if the counting rate is called A, the activity of the sample, Equation 3 can be written in terms of A as well as n, in a relative way, to give

$$\log_{10} \frac{n}{n_0} = \log_{10} \frac{A}{A_0} = -\frac{kt}{2.3} \quad \text{or} \quad \log A = \log A_0 - \frac{kt}{2.3} \qquad (5)$$

To determine the decay constant k for a given active nucleus, we need merely to measure the activity of a sample containing that kind of nucleus as a function of time. Substituting into Equation 5 at two different times and counting rates will permit the elimination of $\log A_0$ and the evaluation of k. A more accurate method would be to measure the activity A of the sample at several times and to make a graph of $\log A$ as a function of time. Since $\log A$ varies linearly with time, the slope of the graph of $\log A$ versus t should be constant and equal to $-k/2.3$.

In this experiment we will use the latter procedure to measure the half-life of a radioactive isotope. Since most naturally occurring radioactive isotopes have very long half-lives, and thus could not be studied by this method, we will use synthetic radioactive isotopes prepared in this laboratory. Many nuclei, when bombarded with slow neutrons, absorb the neutrons and so undergo nuclear reactions. The nuclei so formed are typically radioactive but do not decay by neutron emission. Rather, they emit electrons of high energy and are thereby converted to nuclei of one higher atomic number. Naturally occurring iodine is typical in its behavior in this regard:

On bombardment with slow neutrons: $^{127}_{53}\text{I} + ^{1}_{0}\text{n} \rightarrow ^{128}_{53}\text{I}$

Decay reaction: $^{128}_{53}\text{I} \rightarrow ^{128}_{54}\text{Xe} + ^{0}_{-1}\text{e}$

The electrons, which are called β particles in such reactions, can be readily detected in a counter and so used to measure the decay rate. The isotopes we shall be using have half-lives of the order of a few hours, so that they can be conveniently studied in the course of a laboratory period. The radiation emitted is of relatively low energy and of short duration, and is not particularly hazardous. You should, however, use due caution and not get any sample into your mouth or let it remain for any length of time on your skin.

EXPERIMENTAL PROCEDURE

Obtain a test tube containing an unknown radioactive sample from the stockroom.

Your instructor will demonstrate the use of the radioactivity counter in your laboratory. Do not make any adjustments of the counter settings unless you are specifically directed to do so.

Pour your sample into the planchet provided and smooth it out with the spatula so that it lies in a layer of uniform thickness. Place the planchet in the counter; turn the counter on and measure the number of counts recorded in a one-minute interval. Record the time at which you started the counter. Remove the planchet from the counter and place an empty planchet in the counter. Record the number of background counts obtained in a one-minute interval with the empty planchet.

Wait about 15 min and repeat the procedure, noting the number of counts per minute and the time you begin to count. If there is not too much congestion from other students wishing to use the counter, measure the background counting rate a second time.

Repeat the procedure at about 15-min intervals until you have obtained six counting rates at six different times. During the course of the period you should also obtain about four measurements of the background radiation counting rate.

When you have finished the experiment, dispose of your sample as directed by your laboratory instructor.

DATA AND CALCULATIONS: Determination of the Half-life of a Radioactive Isotope

Unknown no. _____

Count obtained in one minute	Time when counter was turned on	Total elapsed time	Corrected counting rate* = activity, A	$\log_{10}$ corrected counting rate = $\log_{10} A$
_____	_____	_____	_____	_____
_____	_____	_____	_____	_____
_____	_____	_____	_____	_____
_____	_____	_____	_____	_____
_____	_____	_____	_____	_____
_____	_____	_____	_____	_____

Background count _____ _____ _____ _____

Average background _____ cpm obtained in one minute.

*Subtract the average background counting rate from the counts per minute obtained with the sample to obtain the corrected counting rate, or the activity, A, of the sample.

On the graph paper provided, make a graph of $\log_{10} A$ as a function of elapsed time.

Slope of line obtained in the graph $= \dfrac{\Delta \log A}{\Delta t} =$ _____ $=$ _____

Slope of the line $= -\dfrac{k}{2.303}$; $k =$ _____ min^{-1}

By Equation 4, $t_{\frac{1}{2}} = \dfrac{0.693}{k} =$ _____ $=$ _____ min

DETERMINATION OF THE HALF-LIFE OF
A RADIOACTIVE ISOTOPE (Data and Calculations)

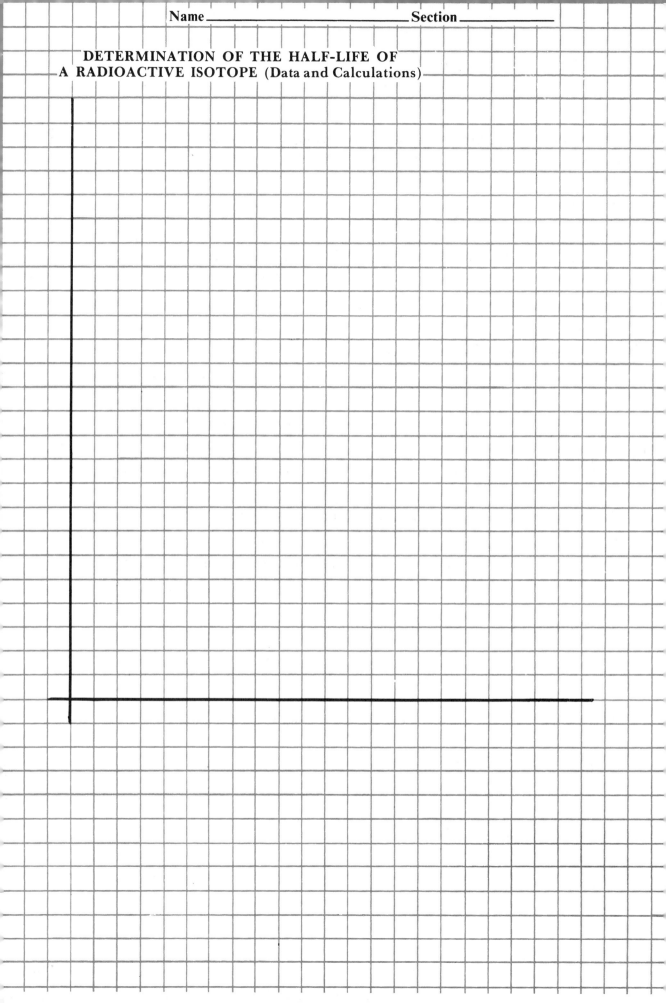

ADVANCE STUDY ASSIGNMENT: **Determination of the Half-life of a**
                                Radioactive Isotope

1. In experiments of this sort the activity ratio A/A_0 can be taken to be equal to the active nuclei ratio n/n_0. Why is this relation valid?

2. A certain sample of a compound containing a radioactive isotope produced 746 counts in one minute at the beginning of the experiment and a count of 428 in one minute when 57 min had passed. If the background count was 20 cpm, what is the decay constant k and the half-life of the radioactive species?

$$k = \text{_____ min}^{-1}$$

$$t_{1/2} = \text{_____ min}$$

QUALITATIVE ANALYSIS

EXPERIMENT

35 • Spot Tests for Some Common Ions

There are two broad categories of problems in analytical chemistry. Quantitative analysis deals with the determination of the amounts of certain species present in a sample; there are several experiments in this manual involving quantitative analysis, and you have probably performed some of them. The other area of analysis, called qualitative analysis, has a more limited purpose, namely, establishing whether given species are or are not present in detectable amounts in a sample. The remaining experiments in this manual will deal mainly with problems in qualitative analysis.

One can carry out the qualitative analysis of a sample in various ways. Probably the simplest approach, which we will use in this experiment, is to test for the presence of each possible component by adding a reagent which will cause that component, if it is in the sample, to react in a characteristic way. This method involves a series of "spot" tests, one for each component, carried out on separate samples of the unknown. The difficulty with this way of doing qualitative analysis is that frequently, particularly in complex mixtures, one species may interfere with the analytical test for another. Although interferences are common, there are many ions which can, under optimum conditions at least, be identified in mixtures by simple spot tests.

In this experiment we will use spot tests for the analysis of a mixture which may contain the following commonly encountered ions in solution:

CO_3^{2-}	PO_4^{3-}	Cl^-	SCN^-
SO_4^{2-}	CrO_4^{2-}	$C_2H_3O_2^-$	NH_4^+

The procedures we will use involve simple acid-base, precipitation, complex ion formation, or oxidation-reduction reactions. In each case you should try to recognize the kind of reaction that occurs, so that you can write the net ionic equation that describes it.

EXPERIMENTAL PROCEDURE

WEAR YOUR SAFETY GLASSES WHILE PERFORMING THIS EXPERIMENT

Carry out the test for each of the anions as directed. Repeat each test using a solution made by diluting the anion solution 9:1 with distilled water; use your 10 cm³ graduated cylinder to make the dilution and make sure you mix well before taking the sample for analysis. In some of the tests, a boiling water bath containing about 100 cm³ water in a 150 cm³ beaker will be needed, so set that up before proceeding. When performing a test, if no reaction is immediately apparent, stir the mixture with your stirring rod to mix the reagents. These tests can easily be used to detect the anions at concentrations of 0.02 M or greater, but in dilute solutions, careful observation may be required.

Test for the Presence of Carbonate Ion, CO_3^{2-}.   Cautiously add 1 cm³ of 6 M HCl to 1 cm³ of 1 M Na_2CO_3 in a small test tube. With concentrated solutions, bubbles of carbon dioxide gas are immediately evolved. With dilute solutions, the effervescence will be much less obvious. Warming in the water bath, with stirring, will increase the amount of bubble formation. Carbon dioxide is colorless and odorless.

Test for the Presence of Sulfate Ion, SO_4^{2-}.   Add 1 cm³ of 6 M HCl to 1 cm³ of 0.5 M Na_2SO_4. Add a few drops of 1 M $BaCl_2$. A white, finely divided precipitate of $BaSO_4$ indicates the presence of SO_4^{2-} ion.

Test for the Presence of Phosphate Ion, PO_4^{3-}.   Add 1 cm³ 6 M HNO_3 to 1 cm³ of 0.5 M Na_2HPO_4. Then add 1 cm³ of 0.5 M $(NH_4)_2MoO_4$ and stir thoroughly. A yellow precipitate of ammonium phosphomolybdate, $(NH_4)_3PO_4 \cdot 12MoO_3$, establishes the presence of phosphate. The precipitate may form slowly, particularly in the more dilute solution; if it does not appear promptly, put the test tube in the boiling water bath for a few minutes.

Test for the Presence of Chromate Ion, CrO_4^{2-}.   Chromate-containing solutions are yellow when neutral or basic and orange under acidic conditions. Chromate ion under highly acidic conditions is a good oxidizing agent and may oxidize such species as SCN^- or NH_4^+. Add 2 cm³ of 6 M HNO_3 to 1 cm³ of 0.5 M K_2CrO_4. If reduction of chromate occurs at this point due to the presence of interfering species, a color change to pale blue will be observed shortly after addition of the acid and may be taken as proof of the presence of chromate ion. If no color change is observed, cool the test tube under the water tap and add 1 cm³ of 3 per cent H_2O_2. If chromate ion is present, a rapidly fading blue color caused by the formation of unstable CrO_5 will be observed.

Test for the Presence of Thiocyanate Ion, SCN^-.   Add 1 cm³ 6 M acetic acid to 1 cm³ of 0.5 M KSCN and stir. Add one or two drops of 0.1 M $Fe(NO_3)_3$. A deep red coloration due to formation of $FeSCN^{2+}$ complex ion is proof of the presence of SCN^- ion.

Test for the Presence of Chloride Ion, Cl^-.   Add 1 cm³ of 6 M HNO_3 to 1 cm³ of 0.5 M NaCl. Add a few drops of 0.1 M $AgNO_3$. A white precipitate of AgCl will form if chloride ion is present.

If thiocyanate ion is present, it will interfere with this test, since it will also form a white precipitate. If the sample contains SCN^- ion, put 1 cm³ of the solution in a small 30 or 50 cm³ beaker and add 1 cm³ of 6 M HNO_3. Boil the solution gently until the volume is decreased to one-half its original value; this will oxidize the thiocyanate and remove the interference. Then add another cm³ of 6 M HNO_3 and a few drops of $AgNO_3$ solution, forming white AgCl in the presence of Cl^- ion as before.

Test for the Presence of Acetate Ion, $C_2H_3O_2^-$.   Add 1 cm³ 3 M H_2SO_4 to 1 cm³ of 1 M $NaC_2H_3O_2$ and stir. If acetate ion is present, acetic acid, with the characteristic odor of vinegar, is formed. The intensity of the odor is enhanced if the tube is warmed for 30 s in the boiling water bath. If the acetate ion concentration is low, boil the neutral solution carefully to near dryness before acidifying it.

Test for the Presence of Ammonium Ion, NH_4^+.   Add 1 cm³ of 6 M NaOH to 1 cm³ of 0.5 M NH_4Cl. If NH_4^+ ion is present, NH_3 will form and can be detected by its characteristic odor. For a more sensitive test, pour the 1 cm³ of sample into a small beaker; moisten a piece of red litmus paper and put it on the bottom of a watch glass; cover the beaker with the watch glass and gently heat the liquid to the boiling point. Do not boil it and be careful that no liquid comes in contact with the litmus paper. If NH_4^+ ion is present, the litmus paper will gradually turn blue as it is exposed to the vapors of NH_3 that will evolve. Remove the watch glass and try to smell the ammonia.

When you have completed all of the tests, obtain an unknown from your laboratory supervisor and analyze it by applying the tests to separate 1 cm³ portions. The unknown will contain 3 or 4 of the ions on the list, so your test for a given ion may be affected by the presence of others. For the most part, the properties of the mixture will be a super-position of the properties of its components, but it is possible that under some conditions the ions will react in unexpected ways. Where a test does not go quite according to the rules, try to deduce why the sample may have behaved as it did. When you think you have properly analyzed your unknown, you may, if you wish, make a "known" which has the composition you found and test it to see if it has the properties of your unknown.

OBSERVATIONS AND REPORT SHEET: Spot Tests for Some Common Ions

Observations and Comments on Spot Tests

Ion	Stock Solution	9:1 Dilution	Unknown
CO_3^{2-}			
SO_4^{2-}			
PO_4^{3-}			
CrO_4^{2-}			
Cl^-			
$C_2H_3O_2^-$			
SCN^-			
NH_4^+			

Unknown No. _____ contains _____

ADVANCE STUDY ASSIGNMENT: Spot Tests for Some Common Ions

1. In this experiment each ion for which we test undergoes a characteristic reaction that serves to establish its presence. Write the net ionic equation for that reaction for each of the ions considered.

CO_3^{2-}

SO_4^{2-}

PO_4^{3-}

CrO_4^{2-}

Cl^-

$C_2H_3O_2^-$

SCN^-

NH_4^+

2. An unknown which might contain any of the ions studied in this experiment has the following properties:

    a. Unknown is yellow and odorless.
    b. On addition of 6 M HNO_3 plus warming, there is no odor.
    c. On addition of 0.1 M $AgNO_3$ to above solution, white precipitate forms.
    d. On addition of 0.1 M $BaCl_2$ to acidified unknown, a precipitate forms.
    e. On addition of 6 M NaOH to unknown, vapor turns moist red litmus paper blue.

On the basis of the above information, which ions are definitely present, which are definitely absent, and which are still in doubt?

         Present   Absent   Doubtful

EXPERIMENT

36 • Identification of Unknown Solids

In the previous experiment you worked with some of the tests that are used to establish the presence of some of the most common anions in a solution. These tests can, in general, be applied to any solution in which the anions may be present and to any solid that is soluble in water.

In this experiment you will be asked to identify, on the basis of their chemical or physical properties, some solid ionic compounds. We have prepared a group of containers, each of which contains a single pure ionic substance. For each container there are two formulas, one of which is that of the substance present in it. Your problem is simply to establish which of the formulas is correct. Your grade will depend on the number of substances you can identify in the time allowed.

Many of the substances can be identified on the basis of their anions, so that you can use the procedures in Experiment 35 to find the correct formula. In other cases the compounds may both be insoluble or may both have the same anion, so that other approaches will be necessary. One very useful procedure is to consider the solubilities of the two compounds in different reagents or the solubilities of products which can be made from those substances.

For example, suppose you were asked to distinguish between $ZnCO_3$ and $BaCO_3$, both of which are insoluble in water. For this problem information regarding the properties of Zn^{2+} and Ba^{2+} would be important, since both of the compounds contain carbonate ion. In Appendix II of this manual is a table summarizing the solubility properties of most of the common cations, including Ba^{2+} and Zn^{2+}; an understanding of the meaning of this table in analyses of this sort can be very helpful. For Ba^{2+} ion the table tells us that $BaCl_2$, $Ba(OH)_2$, and BaS are at least slightly soluble in water. $BaSO_4$ is not soluble in any common reagents; $BaCO_3$, $Ba_3(PO_4)_2$, and $BaCrO_4$ will dissolve in 6 M HCl or 6 M HNO_3. A similar interpretation of the table for Zn^{2+} ion shows that the chloride and sulfate are soluble in water, as is the sulfide under acidic conditions. The carbonate, phosphate, chromate, and hydroxide are all insoluble in water, but will dissolve in 6 M HCl, 6 M NaOH, or 6 M NH_3. Given this information, one could distinguish between $BaCO_3$ and $ZnCO_3$ by any one of the following procedures:

1. Add 6 M HCl to dissolve the solid. To the solution add a few drops of 1 M Na_2SO_4 or 1 M H_2SO_4. A precipitate of $BaSO_4$ indicates the compound was $BaCO_3$. No precipitate means it was $ZnCO_3$.

2. Add 6 M NaOH to the solid. If it dissolves, the solid must have been $ZnCO_3$; the product was $Zn(OH)_4^{2-}$ complex ion. If it is insoluble, $BaCO_3$ is present. Any cation whose insoluble salts will dissolve in a strong base goes into solution as a hydroxo complex.

3. Add 6 M NH_3 to the solid. If it dissolves, the compound has to be $ZnCO_3$. The zinc ammonia complex ion is formed. $BaCO_3$ would remain insoluble. Cations whose insoluble salts go into solution in 6 M NH_3 all form ammonia complex ions.

There are other methods that can be profitably used to identify the solids that you should be aware of. Hydroxides, if they are even slightly soluble, will make their solution basic to litmus or phenolphthalein. Hydrogen sulfate salts, such as $NaHSO_4$, will produce an anion that itself ionizes to produce an acidic solution.

Salts of volatile weak acids, including carbonates, acetates, sulfides, and sulfites,

281

when treated with 6 M H_2SO_4, typically evolve vapors which have distinctive characteristics:

$$CuCO_3(s) + 2\ H^+(aq) \rightarrow CO_2(g) + H_2O + Cu^{2+}(aq) \quad \text{effervescence}$$

$$AgC_2H_3O_2(s) + H^+(aq) \rightarrow HC_2H_3O_2(g) + Ag^+(aq) \quad \text{odor of vinegar}$$

$$ZnS(s) + 2\ H^+(aq) \rightarrow H_2S(g) + Zn^{2+}(aq) \quad \text{odor of rotten eggs}$$

$$K_2SO_3(s) + 2\ H^+(aq) \rightarrow SO_2(g) + H_2O + 2\ K^+(aq) \quad \text{odor of burning sulfur}$$

(Some sulfides are too insoluble to react with 6 M H_2SO_4; see Appendix II, next to last column, where there must be an S if this reaction is to be useful.)

In a few cases, particularly with different halides of the same cation, oxidation-reduction reactions can be useful for characterization. Bromides and iodides, even in very insoluble substances, can be oxidized by chlorine water, forming Br_2 and I_2, which dissolve in trichloroethane to produce orange and violet solutions respectively.

In many cases one or two very simple tests will suffice for the proper identification of an unknown. For each identification you make, write a net ionic equation to describe the reaction on which your decision was based.

EXPERIMENTAL PROCEDURE   WEAR YOUR SAFETY GLASSES WHILE PERFORMING THIS EXPERIMENT

On each laboratory table there will be a set of numbered bottles, each bottle containing a pure solid compound. With each set there is a list on which you will find the number of each bottle and the formulas of two substances the bottle may contain.

When your instructor tells you to begin, select a bottle from the set and write your name on the line alongside the bottle number and its possible contents. Carry out tests to determine which of the two compounds the bottle actually contains. When you are able to decide, write the formula of the compound present on the data sheet, along with the bottle number and the two compounds between which you had to distinguish. On the same line write the net ionic equation for the reaction by which you were able to identify the compound. Go to your instructor, who will tell you if your reasoning is correct. Then proceed to select another bottle and determine its contents. Do as many analyses as you can in the time allowed.

At the end of about half an hour, you will be given another set of unknowns on which to work. You will be allowed to work on a total of four sets of bottles, and you will have about half an hour to spend on each set. For each of the first seven analyses you do correctly, you will obtain one point. For each correct analysis above seven, you will get half a point.

Only one student may work with a given unknown in a given set. A given reaction may be used only *once* for distinguishing between any two compounds; if you use the reaction $Ag^+ + Cl^- \rightarrow AgCl(s)$ to identify a substance, you may not use that reaction again to distinguish between another pair of possibilities.

REPORT SHEET: Identification of Unknown Solids

Unknown Number	Possible Compounds	Compound Present	Identifying reaction (Net ionic equation)
_____	_____	_____	_____
	_____		
_____	_____	_____	_____
	_____		
_____	_____	_____	_____
	_____		
_____	_____	_____	_____
	_____		
_____	_____	_____	_____
	_____		
_____	_____	_____	_____
	_____		
_____	_____	_____	_____
	_____		
_____	_____	_____	_____
	_____		
_____	_____	_____	_____
	_____		
_____	_____	_____	_____
	_____		
_____	_____	_____	_____
	_____		

 Name ______________________ Section ______________

ADVANCE STUDY ASSIGNMENT: Identification of Unknown Solids

1. Devise one- or two-step procedures to determine whether an unknown solid is:

   a. Na_2CO_3 or Na_2SO_4

   b. $CuCl_2$ or $BaCl_2$

   c. KCl or NH_4Cl

   d. NaOH or $NaHSO_4$

   e. $KC_2H_3O_2$ or NH_4NO_3

   f. $NaHSO_4$ or $NaNO_3$

   g. $MgCO_3$ or $MgCl_2$

   h. $AgNO_3$ or $Zn(NO_3)_2$

   i. AgCl or AgBr

   j. CaO or CuO

Qualitative Analysis of Cations

GENERAL PRINCIPLES

In the next group of four experiments we will be concerned with some of the classical procedures that have been devised to test for the presence in solution of some of the more common metallic cations. In each experiment you will study a group of ions which have some common property that allows their separation from other ions. By taking advantage of the differences in properties of the ions within a group, you will be able to separate the ions and check for their presence by confirmatory tests. When you have completed work on a solution containing all the ions in a group, you will be given an unknown to analyze for the presence or absence of those ions. Upon completing your study of the three groups of cations which you will investigate, you will be assigned a set of five ions chosen from these groups. For this set of ions you will devise a scheme of analysis and use it in a laboratory examination to analyze an unknown.

In any qualitative analysis scheme we use several different kinds of chemical reactions. Basic to the scheme are precipitation reactions, by which a group of ions may be separated from other groups, or by which a single ion may be removed and identified by the characteristic nature of its precipitate. Other important kinds of reactions involve the formation of complex ions, acid-base reactions, or oxidation-reduction processes. These reactions are typically used to separate and identify ions within a group. Many precipitation, complex-ion, and oxidation-reduction reactions are markedly influenced by the pH of the solution in which they occur, so that much of the effectiveness of the separation scheme depends on proper control of the acidic or basic nature of the system undergoing reaction.

The first group of cations we will study includes the Ag^+, Hg_2^{2+}, and Pb^{2+} ions. These ions all form insoluble chlorides and comprise Group I in the analysis scheme. They are separated from the other cations included in the scheme by addition of HCl solution to a solution containing all the ions to be investigated. If the HCl is added in excess, AgCl(s), Hg_2Cl_2(s), and $PbCl_2$(s) will be, within the limits noted in Experiments 22 and 23, precipitated quantitatively, and the other cations, which have soluble chlorides, will remain in solution. The Group I precipitate obtained by filtration or decantation can then be tested for the possible presence of Ag^+, Hg_2^{2+}, and Pb^{2+}. It is necessary that the removal of Group I be as quantitative as possible, since the cations in that group will, if present in the solution after the separation step, interfere with the analysis for the cations in other groups.

The Group II cations to be separated are the Cu^{2+}, Bi^{3+}, Sn^{4+}, and Sb^{3+} ions. These four cations all form insoluble sulfides. The are separated from the ions in Group III by precipitation in an acidic sulfide solution.

You may recall that H_2S solutions are very weakly acidic.

$$H_2S(aq) \rightleftharpoons H^+(aq) + HS^-(aq) \qquad K_1 = 1 \times 10^{-7}$$

$$HS^-(aq) \rightleftharpoons H^+(aq) + S^{2-}(aq) \qquad K_2 = 1 \times 10^{-15}$$

$$H_2S(aq) \rightleftharpoons 2\,H^+(aq) + S^{2-}(aq) \qquad K = K_1 \times K_2 = 1 \times 10^{-22}$$

The sulfide ion concentration in H_2S solutions to which a strong acid or base has been added is markedly affected by the acid or base. In a solution saturated with H_2S gas, $[H_2S]$ equals about 0.1 mol/dm³. Therefore, by the last equation, in an H_2S solution whose pH equals 1 as established by the addition of, for example, HCl, $[S^{2-}]$ equals about 1×10^{-20}. In a saturated H_2S solution whose pH is fixed at a value of 9 by the addition of NH_3, $[S^{2-}]$ equals 1×10^{-4}, which is larger than it is in acid solution by a factor of 10^{16}.

The sulfides of the ions in Group II are extremely insoluble (K_{sp} for $CuS = 6 \times 10^{-36}$), so that they will precipitate from mildly acidic H_2S solutions. The sulfides of the Group III ions, Fe^{3+}, Al^{3+}, Cr^{3+}, and Ni^{2+}, are all much more soluble than those of Group II and thus are not precipitated from acidic H_2S solution.

The Group II precipitate is removed from the solution, and the ions are separated into two sub-groups. These are then tested for the presence of the individual ions.

To separate the ions in Group III, Cr^{3+} is first oxidized to CrO_4^{2-}. A concentrated solution of NaOH is added to precipitate $Fe(OH)_3$ and $Ni(OH)_2$. Aluminum, whose hydroxide is amphoteric, remains in solution as $Al(OH)_4^-$. By carefully adjusting the pH of this solution, it is possible to precipitate $Al(OH)_3$; chromium remains in solution as CrO_4^{2-}.

The hydroxides of Fe^{3+} and Ni^{2+} are separated by treatment with NH_3, which brings nickel into solution as the $Ni(NH_3)_6^{2+}$ ion. Since the ammine complex of Fe^{3+} is less stable, $Fe(OH)_3$ does not dissolve in ammonia. The individual ions are then identified by specific tests.

The separation scheme for the 11 ions to be investigated illustrates many of the chemical properties that are associated with cations. In your study of qualitative analysis you should do your best to learn the chemical properties of the substances that appear in the scheme. In order to be able to speak or write sensibly about chemical substances, the chemist has to know something about the ways substances can be made, their reactivities, their colors, their solubilities in various media, and the tests by which they are identified in mixtures. One of the reasons for studying qualitative analysis is to gain such information about a group of typical substances, and you will find that your efforts to this end will be very worthwhile.

The qualitative analysis scheme also offers you a good opportunity to review and apply in logical fashion many, indeed most, of the principles studied during this course. There is nothing in the scheme that a reasonably intelligent student, given time and persistence, could not himself devise. You should make every effort to understand the reason for each step in the procedure. If you know what you are trying to accomplish by a given step, you are much less likely to do something foolish. Try to see how a principle can be applied to a given reaction and what the underlying reason is for this precipitation to occur or that substance to dissolve. If you can get to the point where you can see some of the many relations between the theory and its practical applications in the qualitative analysis scheme, you are well on your way to an understanding of the real significance of the theory.

EXPERIMENT

37 • Qualitative Analysis of Group I Cations

PRECIPITATION AND SEPARATION OF GROUP I IONS

The chlorides of Pb^{2+}, Hg_2^{2+}, and Ag^+ are all insoluble in cold water and relatively dilute chloride media. Thus they can be removed as a group from solution by the addition of HCl until precipitation is essentially complete. It is important to add enough HCl to ensure complete precipitation, but not too large an excess. In concentrated HCl solutions these chlorides tend to dissolve, producing chloro-complexes, such as $AgCl_2^-$.

Lead chloride is separated from the other two insoluble chlorides on the basis of its relative solubility in hot water. Once the Pb^{2+} has been put in solution by extraction of the chlorides with hot water, we can check for its presence with some reagent that gives an identifying reaction with Pb^{2+}. Chromate ion, CrO_4^{2-}, present in a solution of K_2CrO_4, forms a characteristic yellow precipitate with Pb^{2+} in a weakly acidic solution.

The other two insoluble chlorides, AgCl and Hg_2Cl_2, can be separated by the addition of aqueous ammonia. Silver chloride dissolves with the formation of the complex ion $Ag(NH_3)_2^+$. The ammonia also reacts with the mercury(I) chloride to give the precipitate a characteristic gray or black color, which can be used as the confirmation test for the presence of Hg_2^{2+}. The reaction between the mercury(I) chloride and ammonia is a disproportionation reaction in which the mercury(I) is both oxidized and reduced. The principal products of the reaction are mercury(II) amido chloride, $HgNH_2Cl$, and metallic mercury. The presence of the latter in the precipitate turns it black.

The solution containing the $Ag(NH_3)_2^+$ needs to be further tested to establish the presence of silver. The addition of nitric acid to the solution destroys the silver-ammonia complex and reprecipitates white silver chloride.

EXPERIMENTAL PROCEDURE

WEAR YOUR SAFETY GLASSES WHILE PERFORMING THIS EXPERIMENT

1. Precipitation of Group I Ions. Make up 3 cm³ of a "known" solution for Group I by adding to a test tube 1 cm³ samples of solutions containing Ag^+, Hg_2^{2+}, and Pb^{2+}, as obtained by dissolving their nitrate salts.

Add 2 drops of 6 M HCl to 1 cm³ of the solution in a small test tube. Centrifuge the solution, being careful to place a blank test tube containing an equal volume of water in the opposite tube of the centrifuge. Add one more drop of the 6 M HCl to the solution to test for completeness of precipitation. Centrifuge again if necessary, and decant the supernatant solution from the chloride precipitate. The solution should be saved for further study if ions from other groups may be present.

2. Separation of Pb^{2+}. Wash the precipitate with 1 or 2 cm³ of water. Stir with a glass rod, centrifuge, and decant. Discard the decanted liquid.

Add 1 cm³ of distilled water to the precipitate in the test tube and place in a 100 cm³ beaker which is half full of boiling water. Allow the tube to remain in the boiling water bath for a few minutes, and stir occasionally with a glass rod.

Centrifuge the hot solution and quickly pour it into another test tube. Save the remaining precipitate for further study.

3. Identification of Pb^{2+}. Add one drop of 6 M acetic acid and a few drops of 1 M

K_2CrO_4 to the solution from 2. If Pb^{2+} is present in the solution, a yellow precipitate of $PbCrO_4$ will form.

4. Separation and Identification of Hg_2^{2+}. Add 10 drops of 6 *M* NH_3 to the precipitate from 2 and stir thoroughly. Centrifuge the solution and decant. A gray or black precipitate, produced by reaction of Hg_2Cl_2 with ammonia to produce metallic mercury, will establish the presence of Hg_2^{2+}.

5. Identification of Ag^+. Add 6 *M* HNO_3 to the solution from 4 until it is acidic toward litmus paper. Test for acidity by dipping the end of your stirring rod in the solution and then touching it to a piece of litmus paper. If Ag^+ is present in the acidified solution, a white precipitate of $AgCl$ will form.

6. When you have completed the tests on the known solution, obtain an unknown and analyze it for the possible presence of Ag^+, Pb^{2+}, and Hg_2^{2+}.

FLOW DIAGRAMS

It is possible to summarize the directions for analysis of the Group I cations in what is called a flow diagram. In the diagram, successive steps in the procedure are linked with arrows. Reactant cations or reactant substances containing the ions are at one end of each arrow and products formed are at the other end. Reagents and conditions used to carry out each step are placed alongside the arrows. A partially completed flow diagram for the Group I ions follows:

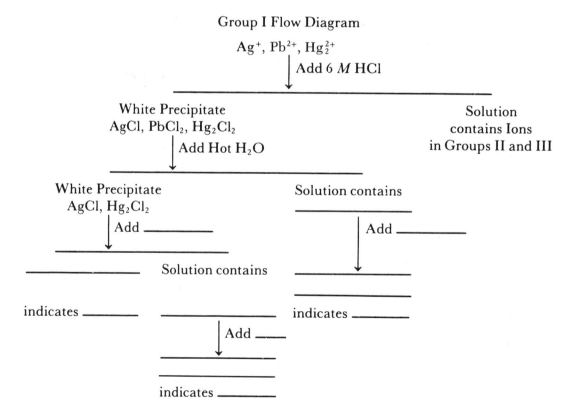

Group I Flow Diagram

You will find it useful to construct flow diagrams for each of the cation groups. You can use such diagrams in the laboratory to serve as a brief guide to procedure, and you can use them to directly record your observations on your known and unknown solutions.

Name _____ Section _____

**OBSERVATIONS AND REPORT SHEET: Qualitative Analysis of
Group I Cations**

Flow Diagram for Group I

Observations on known (record on diagram if different from those on prepared diagram).

Observations on unknown (record on diagram in colored pencil to distinguish from observations on known).

Unknown no. _____

Ions reported present _____ _____ _____

291

ADVANCE STUDY ASSIGNMENT: Group I Cations

1. On the report sheet, complete the flow diagram for the separation and identification of the ions in Group I.

2. Write balanced net ionic equations for the following reactions:

   a. The precipitation of the chloride of Pb^{2+}.

   b. The confirmatory test for Hg_2^{2+}.

   c. The dissolving of AgCl in aqueous ammonia.

   d. The precipitation of AgCl from the solution of $Ag(NH_3)_2^+$.

3. A solution may contain Ag^+, Pb^{2+}, and Hg_2^{2+}. A white precipitate forms on addition of 6 M HCl. The precipitate is insoluble in hot water and dissolves on addition of ammonia. Which of the ions are present, which are absent, and which remain undetermined? State your reasoning.

Present _____

Absent _____

Doubtful _____   **293**

EXPERIMENT

38 • Qualitative Analysis of Group II Cations

PRECIPITATION AND SEPARATION OF THE GROUP II IONS

The sulfides of the four ions in Group II, Bi^{3+}, Sn^{4+}, Sb^{3+}, and Cu^{2+}, are insoluble at a pH of 0.5. The solution is adjusted to this pH and then saturated with H_2S, which may be done by simply bubbling H_2S from a generator through the solution. A more convenient method, however, is to boil the acid solution after adding a small amount of thioacetamide. This compound, CH_3CSNH_2, hydrolyzes when heated in water solution to liberate H_2S:

$$CH_3CSNH_2(aq) + 2\,H_2O \rightleftharpoons H_2S(aq) + CH_3COO^-(aq) + NH_4^+(aq)$$

Using thioacetamide as the precipitating reagent has the advantage of minimizing odor problems and giving denser precipitates.

The four insoluble sulfides can be separated into two groups by extraction with an alkaline solution containing sulfide ion. The sulfides of tin and antimony dissolve, forming the anionic hydroxo-complexes, $Sn(OH)_6^{2-}$ and $Sb(OH)_4^-$, or anionic thio-complexes, SnS_3^{2-} and SbS_3^{3-}. The solution containing the ions of Sb and Sn is reacidified with HCl and treated again with thioacetamide to precipitate the sulfides. The Sb_2S_3 and SnS_2 are then dissolved as chloro-complexes in concentrated HCl, and their presence is confirmed by appropriate tests.

The confirmatory test for tin takes advantage of the two oxidation states, +2 and +4, in which tin can exist in aqueous solution. Aluminum is added to the acid solution containing Sn^{4+} to reduce it to Sn^{2+}. The addition of $HgCl_2$ to this solution will cause another oxidation-reduction reaction and produce a gray or black precipitate containing Hg or Hg_2Cl_2.

If oxalic acid is added to the solution containing the chloro-complexes of Sb^{3+} and Sn^{4+}, a very stable oxalato-complex of tin is formed, in the presence of which the bright orange Sb_2S_3 can be precipitated to establish the presence of Sb^{3+}.

The remaining two sulfides, CuS and Bi_2S_3, which are not soluble in alkaline sulfide solution, will dissolve in a strongly oxidizing acid such as HNO_3. The nitric acid oxidizes the sulfur in these sulfides to the free element or up to the sulfate oxidation state with prolonged heating. The cations go into solution without change in oxidation state. The Cu^{2+} and the Bi^{3+} can easily be separated by the addition of aqueous ammonia. Bismuth (III) hydroxide precipitates and the copper stays in solution as the deep blue $Cu(NH_3)_4^{2+}$ complex ion.

The bismuth hydroxide is dissolved by treating with hydrochloric acid. The solution is poured into distilled water. If Bi^{3+} is present a white precipitate of BiOCl will form.

EXPERIMENTAL PROCEDURE

1. **Adjustment of pH Prior to Precipitation.** Obtain a 1 cm³ sample of the "known" solution for Group II, containing 0.1 M concentrations of the nitrates or chlorides of

295

Sn^{4+}, Sb^{3+}, Cu^{2+}, and Bi^{3+}. Add 15 M NH_3 until the solution is just basic to litmus paper. A precipitate will probably form during this step because of formation of insoluble oxychlorides of bismuth, tin, or antimony. Then add 1 drop of 6 M HCl for each 1 cm^3 of solution. This should bring the solution to a pH of about 0.5. This can be verified by the use of methyl violet indicator paper with 0.3 M HCl solution (pH = 0.5) as a standard for comparison. This paper can be prepared by placing methyl violet solution on filter paper and allowing it to dry. The 0.3 M HCl solution will give the methyl violet spot a blue-green color. Adjust the pH of your test solution if necessary by adding HCl or NH_3 so that it gives the same color.

2. Precipitation of Group II Sulfides. Add 15 to 20 drops of 1 M thioacetamide solution to your test solution and heat it in a boiling water bath for 5 min. Centrifuge the solution, which will contain the Group III cations if they are present, and decant into a second test tube. Test the solution for completeness of precipitation by adding 2 drops of thioacetamide and allowing it to stand for 1 min. If a precipitate forms, add a few more drops of thioacetamide solution and heat again in the boiling water bath. Combine the two batches of precipitate. Wash the precipitate with 1 M NH_4Cl solution, stir thoroughly and warm in the water bath; centrifuge, and discard the wash solution.

3. Separation of Group II Sulfides into Two Subgroups. To the sulfide precipitate add 15 drops of 6 M KOH, 2 cm^3 of water, and 3 drops of thioacetamide solution. Heat the solution in a boiling water bath for 5 min. Stir the solution occasionally. Centrifuge and decant the solution immediately after heating. Wash any remaining precipitate twice with 5 drops of water and combine the washings with the solution. At this point tin and antimony will be present in the solution as complex ions, and the copper and bismuth will remain as an undissolved sulfide precipitate.

4. Reprecipitation of SnS_2 and Sb_2S_3. Add 12 M HCl to the solution from 3 until it is just acidic toward litmus paper. Upon acidification, some of the tin and antimony will again precipitate as orange sulfides. Add 5 drops of thioacetamide solution and heat as before to complete the precipitation. Centrifuge and decant the solution. The solution may be discarded.

5. Dissolving SnS_2 and Sb_2S_3 in Acid Solution. Add 10 to 15 drops of 12 M HCl to the precipitate from 4 and heat in a boiling water bath for 5 min to dissolve the precipitate. Heat the test tube in an air bath (an empty beaker) for a few minutes, boiling gently to drive out the H_2S. Centrifuge out any insoluble sulfur residue.

6. Confirmation of the Presence of Tin. Test the solution for tin by adding a piece of aluminum wire and 1 cm^3 of 12 M HCl to about half of the solution from 5. The aluminum will reduce any Sn^{4+} present to Sn^{2+}. Heat the solution in the boiling water bath for 5 min, centrifuge, and decant. A black residue is indicative of the presence of antimony. Add 2 to 3 drops of 0.1 M $HgCl_2$ to the solution. This will cause an oxidation-reduction reaction, giving a white or gray precipitate of Hg_2Cl_2 or Hg. The formation of this precipitate confirms the presence of tin.

7. Test for Antimony. Add 0.5 g of oxalic acid and 5 ml of water to the remainder of the solution from step 5. Heat the solution, if necessary, to dissolve the oxalic acid. This reagent forms a very stable complex ion with the Sn^{4+}. Add 10 drops of thioacetamide solution and heat as before in a boiling water bath. The formation of an orange precipitate, Sb_2S_3, confirms the presence of antimony.

8. Dissolving CuS and Bi_2S_3. Add 30 drops of 6 M HNO_3 to the remaining sulfide precipitate from step 3. Heat the solution gently in the water bath to complete the dissolution of the sulfides. Centrifuge the solution to remove sulfur and decant it into a small beaker. Evaporate to a volume of about 0.5 cm^3.

9. Separation of Cu²⁺ and Bi³⁺ and Identification of Cu²⁺. Add 15 M NH₃ dropwise to the solution obtained in step 8 until it is just basic toward litmus paper. Add a few drops more. The formation of a deep blue solution characteristic of the $Cu(NH_3)_4^{2+}$ ion establishes the presence of copper.

10. Test for Bismuth. Any white precipitate formed upon the addition of the NH₃ in step 9 is probably $Bi(OH)_3$. To test for the presence of Bi^{3+}, centrifuge and decant the solution in 9. Dissolve the precipitate in 5 drops 6 M HCl. Centrifuge out any undissolved sulfur residue. Pour the solution into a 600 cm³ beaker two-thirds full of distilled water at room temperature. A white precipitate of BiOCl confirms the presence of bismuth.

11. When you have completed the analysis of your known, obtain a Group II unknown and test it for the possible presence of Cu^{2+}, Bi^{3+}, Sn^{4+}, and Sb^{3+}.

OBSERVATIONS AND REPORT SHEET: Qualitative Analysis of
Group II Cations

Flow Diagram for Group II

Observations on known (record on diagram if different from those on prepared diagram).

Observations on unknown (record on diagram in colored pencil to distinguish from observations on known).

Unknown no. _____

Ions reported present _____ _____ _____ _____

ADVANCE STUDY ASSIGNMENT: Qualitative Analysis of
                             Group II Cations

1. Prepare a flow diagram for the separation and identification of the Group II ions and put it on your report sheet.

2. Write balanced net ionic equations for the following reactions:

    a. Precipitation of the bismuth sulfide with H_2S.

    b. The confirmatory test for copper.

    c. The confirmatory test for bismuth.

    d. The dissolving of CuS in hot nitric acid.

3. A solution which may contain Cu^{2+}, Bi^{3+}, Sn^{4+}, or Sb^{3+} ions is treated with thio-acetamide in an acid medium. The black precipitate that forms is partly soluble in strongly alkaline solution. The precipitate that remains is soluble in $6\,M$ HNO_3 and gives only a deep blue solution on treatment with excess NH_3. The alkaline solution, when acidified, produces an orange precipitate. On the basis of this information, which ions are present, which are absent, and which are still in doubt?

                                        Present _____

                                        Absent   _____

                                        In doubt _____

EXPERIMENT
39 • Qualitative Analysis of Group III Cations

PRECIPITATION AND SEPARATION OF THE GROUP III IONS

The four ions in group III are Cr^{3+}, Al^{3+}, Fe^{3+}, and Ni^{2+}. The first step in the analysis involves treating the solution with sodium hydroxide, adding hydrogen peroxide, and heating. Under these conditions, chromium(III) is oxidized to CrO_4^{2-}. The chromate ion remains in solution along with the hydroxo-complex formed by aluminum(III), $Al(OH)_4^-$. On the other hand, iron(III) and nickel(II) form precipitates of $Fe(OH)_3$ and $Ni(OH)_2$ respectively. The hydroxides of these ions, unlike $Al(OH)_3$, are not amphoteric and so precipitate even in strongly basic solution.

The Cr(VI) is separated from Al(III) by acidification and then precipitation of $Al(OH)_3$ with ammonia. The CrO_4^{2-} is precipitated as yellow $BaCrO_4$ by the addition of $BaCl_2$. The precipitate of $BaCrO_4$ is dissolved in HNO_3 to give orange $Cr_2O_7^{2-}$ in solution. A deep blue color is produced when H_2O_2 is added to this solution. This color, which is due to the presence of a peroxo-compound, probably CrO_5, is used to confirm the presence of chromium in the sample. If diethyl ether is added, it extracts the blue-colored species. We can confirm the presence of aluminum in the sample by dissolving the gelatinous hydroxide and adding aluminon reagent to the reprecipitated hydroxide to form a red lake.

The precipitate containing the $Ni(OH)_2$ and $Fe(OH)_3$ is dissolved in HCl. The addition of NH_3 reprecipitates $Fe(OH)_3$ and leaves the nickel in solution as $Ni(NH_3)_6^{2+}$. The test for nickel in the solution is made by adding dimethylglyoxime, $C_4H_8N_2O_2$ (H_2DMG). This organic precipitating agent gives a deep rose-colored precipitate in the presence of nickel, $NiC_8H_{14}N_4O_4$ ($Ni(HDMG)_2$). We can determine the presence of iron by dissolving the precipitate of $Fe(OH)_3$ in HCl and adding KSCN. If iron(III) is present in the solution, blood-red $FeSCN^{2+}$ will form.

EXPERIMENTAL PROCEDURE

WEAR YOUR SAFETY GLASSES WHILE PERFORMING THIS EXPERIMENT

1. If you are testing a solution from which Group II cations have been precipitated (Experiment 38), remove the H_2S and excess acid by boiling the solution in a small beaker until the volume is reduced to about 1 cm^3. Remove any sulfur residue by centrifuging the solution.

If you are working on the analysis of Group III ions only, prepare a known solution containing Fe^{3+}, Al^{3+}, Cr^{3+}, and Ni^{2+} by mixing together 1 cm^3 portions of each of the appropriate 0.1 M solutions.

2. **Oxidation of Cr(III) to Cr(VI) and Separation of the Insoluble Hydroxides.** Add 3 drops of 3 per cent H_2O_2 to the solution. Add 15 drops of 6 M NaOH to 1 cm^3 of the solution in a small test tube, making it strongly alkaline. Stir for one minute and *boil carefully* to remove all excess H_2O_2; when the H_2O_2 has been decomposed, the solution will tend to "bump." If the oxidation of Cr(III) has occurred, the solution should be yellow. If it is not, add 3 more drops H_2O_2 and carefully boil again. Centrifuge and decant. Save both the precipitate, which contains iron and nickel hydroxides, and the solution, which contains chromium and aluminum (CrO_4^{2-} and $Al(OH)_4^-$) ions.

3.   Separation of Al from Cr and the Detection of Aluminum.   Acidify the solution from 2 by adding drops of 15 M HNO$_3$ (test with litmus). Add 6 M NH$_3$ drop by drop until the solution is basic. This treatment will bring down Al^{3+} as Al(OH)$_3$, a characteristic gelatinous white precipitate. The precipitate may appear yellow because of the presence of CrO$_4^{2-}$. Centrifuge and decant the solution, which contains CrO$_4^{2-}$. Wash the precipitate with 1 cm^3 hot water; discard the wash liquid. Dissolve the precipitate in a few drops of 6 M HNO$_3$ and centrifuge out any insoluble residue. To the decantate add 2 drops of aluminon solution. Mix well and make barely alkaline with 6 M NH$_3$. A cherry-red precipitate due to a lake of Al(OH)$_3$ and adsorbed aluminon proves the presence of aluminum.

4.   Confirmation of Chromium.   Add 1 M BaCl$_2$ to the solution from step 3 to precipitate yellow BaCrO$_4$. If the precipitate is slow to form, heat the test tube in the boiling water bath. Centrifuge the solution and decant. Dissolve the precipitate in 3 drops of 6 M HNO$_3$; heat gently and stir for about a minute. Add 4 drops of water and cool the test tube in cold water. Add 1 drop of 3 per cent H$_2$O$_2$ and 2 drops of diethyl ether. Shake the tube gently and allow the layers to settle. A blue color in either layer, which may quickly fade, is the confirmatory test for chromium.

5.   Separation of Ni^{2+} and Fe^{3+}.   Add 8 drops of 15 M HNO$_3$ to the precipitate from 2. Heat in the boiling water bath until it is completely dissolved. Allow to cool and add 10 drops of 1 M NH$_4$Cl and 15 M NH$_3$ until the solution is alkaline to litmus paper. This will reprecipitate the iron as brown Fe(OH)$_3$. Add 4 to 5 drops more of the NH$_3$ solution and stir to bring all of the nickel into solution as Ni(NH$_3$)$_6^{2+}$. Centrifuge and decant the solution, saving both solution and precipitate.

6.   Confirmation of Ni^{2+}.   Add 4 drops of dimethylglyoxime solution to the solution from 5. A red precipitate indicates the presence of nickel.

7.   Confirmation of Fe^{3+}.   Dissolve the precipitate from 5 in 0.5 cm^3 of 6 M HCl. Dilute with 2 cm^3 of water and add 2 drops of 0.5 M KSCN. A blood-red color due to the formation of FeSCN^{2+} establishes the presence of Fe^{3+}.

**8.   **When you have completed analysis of the known solution, obtain a Group III unknown and test it for the possible presence of Al^{3+}, Cr^{3+}, Fe^{3+}, and Ni^{2+}.

OBSERVATIONS AND REPORT SHEET: Qualitative Analysis of
Group III Cations

Flow Diagram for Group III

Observation on known (record on diagram if different from those on prepared diagram)

Observations on unknown (record on diagram in colored pencil to distinguish from observations on known)

Unknown no. _____

Ions reported present _____ _____ _____ _____ **305**

**ADVANCE STUDY ASSIGNMENT: Qualitative Analysis of
                                Group III Cations**

1. Prepare a flow diagram for the separation and identification of the ions in Group III and put it on your report sheet.

2. Write balanced net ionic equations for the following reactions:

   a. Dissolving of $Fe(OH)_3$ in hydrochloric acid.

   b. Oxidation of Cr^{3+} to CrO_4^{2-} by H_2O_2 in alkaline solution.

   c. The confirmatory test for Ni^{2+}.

   d. The solution of $Al(OH)_3$ in excess OH^-.

3.   A solution may contain any of the Group III cations. Treatment of the solution with H_2O_2 in alkaline medium yields a yellow solution and a colored precipitate. The acidified solution produces a gelatinous white precipitate on treatment with NH_3. The colored precipitate dissolves in nitric acid; addition of excess NH_3 to this acid solution produces only a blue solution. On the basis of this information, which Group III cations are present, absent, or doubtful?

                                    Present _____

                                    Absent _____

                                    Doubtful _____

EXPERIMENT

40 • Laboratory Examination on Qualitative Analysis of Cations

In Experiments 37 to 39 you have been studying a procedure for the qualitative analysis of eleven common cations. By successive application of the procedures used for Groups I, II, and III, you could readily analyze a general unknown for the presence of the eleven ions.

Rather than give you a general unknown, we will give you a more limited unknown, one confined to five possible ions from the group of eleven. Each student will be given a different set of ions to work with and will be assigned his set of ions the week previous to this examination; that week he will be given a set of five letters, corresponding to the five cations which may be in his unknown, according to the following code:

$$A = Ag^+ \quad C = Hg_2^{2+} \quad E = Bi^{3+} \quad G = Sb^{3+} \quad I = Al^{3+} \quad K = Fe^{3+}$$
$$B = Pb^{2+} \quad D = Cu^{2+} \quad F = Sn^{4+} \quad H = Ni^{2+} \quad J = Cr^{3+}$$

Before coming to laboratory for the examination, prepare a flow diagram for the analysis of your five ions. During the examination you will analyze your unknown according to your analysis scheme and will report your results on the page with your flow diagram.

You will be allowed 50 min in the laboratory period to complete the analysis. You will receive a bonus if you finish your unknown within 30 min and a penalty if it takes you more than 50 min. No results will be accepted after 75 min.

Your grade on the examination will be based on (1) the accuracy of your analysis, (2) the workability of the scheme for analysis on your flow diagram, (3) the absence of extra steps in your flow sheet, steps that relate to ions other than those that can be in your unknown, and (4) the time it takes you to complete the analysis.

There is to be no communication between students during the examination. During the laboratory period, your laboratory supervisor will not answer any questions concerning the analysis. In preparing your flow diagram you may use the portions of the procedure for Groups I, II, and III, which pertain to the cations that may be in your unknown, or you may use other reactions that you are sure will enable you to make the analysis. It is to your interest to shorten the scheme as much as you can, avoiding any steps that are unnecessary. It is of course important that the scheme you decide upon is one that will work for the set of ions in your unknown.

In developing your analysis scheme there are several precautions you should observe, which are perhaps not at once obvious. Your procedure will in all probability involve removing from the solution first one group of ions, then another, and then perhaps a third. This means that if you are to avoid difficulty in interpretation, the group separation must be *complete*: always test to make sure that the precipitating reagent has been added in sufficient amount to bring down all of the material that should precipitate at that point. If all of the Group I ions are not removed by the addition of an adequate amount of HCl, you may be sure that the remaining Group I ions will precipitate along with Group II when that group is removed from the system.

It is also essential to recognize that when groups of ions are separated by means of a selective precipitation, decantation after centrifuging will leave some of the solution along with the solid in the test tube. The Group I precipitate may be washed with a few drops of cold water to remove ions in Groups II and III. The Group II precipitate may be washed with 1 M NH_4Cl solution to remove Group III cations. The wash liquid should be thoroughly mixed with the solid to dilute the undesired cations. Centrifuge, decant, and discard the wash liquid.

Lead chloride is relatively soluble, so that although most of the Pb^{2+} will be removed in Group I, some PbS may precipitate in Group II. It will be dissolved along with the Bi^{3+} and Cu^{2+} in nitric acid, and will precipitate along with $Bi(OH)_3$. It will not, however, interfere with the test for Bi^{3+} that we have been using.

When carrying out the analysis of your unknown during the examination, be sure to wear your safety glasses and to observe any precautions indicated in the procedures for Groups I, II, and III that pertain to cations that may be present in your sample.

Name _____ Section _____

ADVANCE STUDY ASSIGNMENT
AND LABORATORY REPORT: **Laboratory Examination**

Unknown no. _____

Possible Ions _____ _____ _____ _____ _____

Flow Diagram

Ions found in unknown _____ _____ _____ _____ _____

Time turned in (to be entered by laboratory supervisor)   _____

APPENDIX I

Vapor Pressure of Water

Temperature °C	Pressure kPa	Temperature °C	Pressure kPa
0	0.6	26	3.4
1	0.7	27	3.6
2	0.7	28	3.8
3	0.8	29	4.0
4	0.8	30	4.2
5	0.9	31	4.5
6	0.9	32	4.8
7	1.0	33	5.0
8	1.1	34	5.3
9	1.1	35	5.6
10	1.2	40	7.4
11	1.3	45	9.6
12	1.4	50	12.3
13	1.5	55	15.7
14	1.6	60	19.9
15	1.7	65	25.0
16	1.8	70	31.2
17	1.9	75	38.5
18	2.1	80	47.3
19	2.2	85	57.8
20	2.3	90	70.1
21	2.5	95	84.5
22	2.6	97	90.9
23	2.8	99	97.7
24	3.0	100	101.3
25	3.2	101	105.0

APPENDIX II

Summary of Solubility Properties of Ions and Solids

	Cl^-	SO_4^{2-}	CO_3^{2-}, PO_4^{3-}	CrO_4^{2-}	OH^- O^{2-}	H_2S, pH = 0.5	S^{2-}, pH = 9
Na^+, K^+, NH_4^+	S	S	S	S	S	S	S
Ba^{2+}	S	I	A	A	S^-	S	S
Ca^{2+}	S	S^-	A	S	S^-	S	S
Mg^{2+}	S	S	A	S	A	S	S
Fe^{3+}	S	S	A	A	A	S	A
Cr^{3+}	S	S	A	A	A	S	A
Al^{3+}	S	S	A, B	A, B	A, B	S	A, B
Ni^{2+}	S	S	A, N	A, N	A, N	S	A^+, O^+
Co^{2+}	S	S	A	A	A	S	A^+, O^+
Zn^{2+}	S	S	A, B, N	A, B, N	A, B, N	S	A
Mn^{2+}	S	S	A	A	A	S	A
Cu^{2+}	S	S	A, N	A, N	A, N	O	O
Cd^{2+}	S	S	A, N	A, N	A, N	A^+, O	A^+, O
Bi^{3+}	A	A	A	A	A	O	O
Hg^{2+}	S	S	A	A	A	O^+, C	O^+, C
Sn^{2+}, Sn^{4+}	A, B	A, B	A, B	A, B	A, B	A^+, C	A^+, C
Sb^{3+}	A, B	A, B	A, B	A, B	A, B	A^+, C	A^+, C
Ag^+	A^+, N	S^-, N	A, N	A, N	A, N	O	O
Pb^{2+}	HW, B, A^+	B	A, B	B	A, B	O	O
Hg_2^{2+}	O^+	S^-, A	A	A	A	O	O^+

Key: S, soluble in water.
    A, soluble in acid (6M HCl or other nonprecipitating, non-oxidizing acid).
    B, soluble in 6M NaOH.
    O, soluble in hot 6M HNO$_3$.
    N, soluble in 6M NH$_3$.
    I, insoluble in any common reagent.

S^-, slightly soluble in water.
A^+, soluble in 12M HCl.
O^+, soluble in aqua regia.
C, soluble in 6M NaOH containing excess S^{2-}.
HW, soluble in hot water.

Example: For Cu^{2+} and OH^- the entry is A, N. This means that $Cu(OH)_2(s)$, the product obtained when solutions containing Cu^{2+} and OH^- are mixed, will dissolve to the extent of at least 0.1 mol/dm³ when treated with 6M HCl or 6M NH$_3$. Since 6M HNO$_3$, 12M HCl, and aqua regia are at least as strongly acidic as 6M HCl, $Cu(OH)_2(s)$ would also be soluble in those reagents.

APPENDIX III

Table of Atomic Masses

(Based on Carbon-12)

	Symbol	Atomic No.	Atomic Mass		Symbol	Atomic No.	Atomic Mass
Actinium	Ac	89	[227]*	Mercury	Hg	80	200.59
Aluminum	Al	13	26.9815	Molybdenum	Mo	42	95.94
Americium	Am	95	[243]	Neodymium	Nd	60	144.24
Antimony	Sb	51	121.75	Neon	Ne	10	20.183
Argon	Ar	18	39.948	Neptunium	Np	93	[237]
Arsenic	As	33	74.9216	Nickel	Ni	28	58.71
Astatine	At	85	[210]	Niobium	Nb	41	92.906
Barium	Ba	56	137.34	Nitrogen	N	7	14.0067
Berkelium	Bk	97	[247]	Nobelium	No	102	[253]
Beryllium	Be	4	9.0122	Osmium	Os	76	190.2
Bismuth	Bi	83	208.980	Oxygen	O	8	15.9994
Boron	B	5	10.811	Palladium	Pd	46	106.4
Bromine	Br	35	79.909	Phosphorus	P	15	30.9738
Cadmium	Cd	48	112.40	Platinum	Pt	78	195.09
Calcium	Ca	20	40.08	Plutonium	Pu	94	[242]
Californium	Cf	98	[249]	Polonium	Po	84	[210]
Carbon	C	6	12.01115	Potassium	K	19	39.102
Cerium	Ce	58	140.12	Praseodymium	Pr	59	140.907
Cesium	Cs	55	132.905	Promethium	Pm	61	[145]
Chlorine	Cl	17	35.453	Protactinium	Pa	91	[231]
Chromium	Cr	24	51.996	Radium	Ra	88	[226]
Cobalt	Co	27	58.9332	Radon	Rn	86	[222]
Copper	Cu	29	63.546	Rhenium	Re	75	186.2
Curium	Cm	96	[247]	Rhodium	Rh	45	102.905
Dysprosium	Dy	66	162.50	Rubidium	Rb	37	85.47
Einsteinium	Es	99	[254]	Ruthenium	Ru	44	101.07
Erbium	Er	68	167.26	Samarium	Sm	62	150.35
Europium	Eu	63	151.96	Scandium	Sc	21	44.956
Fermium	Fm	100	[253]	Selenium	Se	34	78.96
Fluorine	F	9	18.9984	Silicon	Si	14	28.086
Francium	Fr	87	[223]	Silver	Ag	47	107.870
Gadolinium	Gd	64	157.25	Sodium	Na	11	22.9898
Gallium	Ga	31	69.72	Strontium	Sr	38	87.62
Germanium	Ge	32	72.59	Sulfur	S	16	32.064
Gold	Au	79	196.967	Tantalum	Ta	73	180.948
Hafnium	Hf	72	178.49	Technetium	Tc	43	[99]
Helium	He	2	4.0026	Tellurium	Te	52	127.60
Holmium	Ho	67	164.930	Terbium	Tb	65	158.924
Hydrogen	H	1	1.00797	Thallium	Tl	81	204.37
Indium	In	49	114.82	Thorium	Th	90	232.038
Iodine	I	53	126.9044	Thulium	Tm	69	168.934
Iridium	Ir	77	192.2	Tin	Sn	50	118.69
Iron	Fe	26	55.847	Titanium	Ti	22	47.90
Krypton	Kr	36	83.80	Tungsten	W	74	183.85
Lanthanum	La	57	138.91	Uranium	U	92	238.03
Lawrencium	Lw	103	[257]	Vanadium	V	23	50.942
Lead	Pb	82	207.19	Xenon	Xe	54	131.30
Lithium	Li	3	6.939	Ytterbium	Yb	70	173.04
Lutetium	Lu	71	174.97	Yttrium	Y	39	88.905
Magnesium	Mg	12	24.312	Zinc	Zn	30	65.37
Manganese	Mn	25	54.9380	Zirconium	Zr	40	91 22
Mendelevium	Md	101	[256]				

*A value given in brackets denotes the mass number of the longest-lived or best-known isotope.

APPENDIX IV

Table of Logarithms

	0	1	2	3	4	5	6	7	8	9
1.0	.0000	.0043	.0086	.0128	.0170	.0212	.0253	.0294	.0334	.0374
1.1	.0414	.0453	.0492	.0531	.0569	.0607	.0645	.0682	.0719	.0755
1.2	.0792	.0828	.0864	.0899	.0934	.0969	.1004	.1038	.1072	.1106
1.3	.1139	.1173	.1206	.1239	.1271	.1303	.1335	.1367	.1399	.1430
1.4	.1461	.1492	.1523	.1553	.1584	.1614	.1644	.1673	.1703	.1732
1.5	.1761	.1790	.1818	.1847	.1875	.1903	.1931	.1959	.1987	.2014
1.6	.2041	.2068	.2095	.2122	.2148	.2175	.2201	.2227	.2253	.2279
1.7	.2304	.2330	.2355	.2380	.2405	.2430	.2455	.2480	.2504	.2529
1.8	.2553	.2577	.2601	.2625	.2648	.2672	.2695	.2718	.2742	.2765
1.9	.2788	.2810	.2833	.2856	.2878	.2900	.2923	.2945	.2967	.2989
2.0	.3010	.3032	.3054	.3075	.3096	.3118	.3139	.3160	.3181	.3201
2.1	.3222	.3243	.3263	.3284	.3304	.3324	.3345	.3365	.3385	.3404
2.2	.3424	.3444	.3464	.3483	.3502	.3522	.3541	.3560	.3579	.3598
2.3	.3617	.3636	.3655	.3674	.3692	.3711	.3729	.3747	.3766	.3784
2.4	.3802	.3820	.3838	.3856	.3874	.3892	.3909	.3927	.3945	.3962
2.5	.3979	.3997	.4014	.4031	.4048	.4065	.4082	.4099	.4116	.4133
2.6	.4150	.4166	.4183	.4200	.4216	.4232	.4249	.4265	.4281	.4298
2.7	.4314	.4330	.4346	.4362	.4378	.4393	.4409	.4425	.4440	.4456
2.8	.4472	.4487	.4502	.4518	.4533	.4548	.4564	.4579	.4594	.4609
2.9	.4624	.4639	.4654	.4669	.4683	.4698	.4713	.4728	.4742	.4757
3.0	.4771	.4786	.4800	.4814	.4829	.4843	.4857	.4871	.4886	.4900
3.1	.4914	.4928	.4942	.4955	.4969	.4983	.4997	.5011	.5024	.5038
3.2	.5051	.5065	.5079	.5092	.5105	.5119	.5132	.5145	.5159	.5172
3.3	.5185	.5198	.5211	.5224	.5237	.5250	.5263	.5276	.5289	.5302
3.4	.5315	.5328	.5340	.5353	.5366	.5378	.5391	.5403	.5416	.5428
3.5	.5441	.5453	.5465	.5478	.5490	.5502	.5514	.5527	.5539	.5551
3.6	.5563	.5575	.5587	.5599	.5611	.5623	.5635	.5647	.5658	.5670
3.7	.5682	.5694	.5705	.5717	.5729	.5740	.5752	.5763	.5775	.5786
3.8	.5798	.5809	.5821	.5832	.5843	.5855	.5866	.5877	.5888	.5899
3.9	.5911	.5922	.5933	.5944	.5955	.5966	.5977	.5988	.5999	.6010
4.0	.6021	.6031	.6042	.6053	.6064	.6075	.6085	.6096	.6107	.6117
4.1	.6128	.6138	.6149	.6160	.6170	.6180	.6191	.6201	.6212	.6222
4.2	.6232	.6243	.6253	.6263	.6274	.6284	.6294	.6304	.6314	.6325
4.3	.6335	.6345	.6355	.6365	.6375	.6385	.6395	.6405	.6415	.6425
4.4	.6435	.6444	.6454	.6464	.6474	.6484	.6493	.6503	.6513	.6522
4.5	.6532	.6542	.6551	.6561	.6571	.6580	.6590	.6599	.6609	.6618
4.6	.6628	.6637	.6646	.6656	.6665	.6675	.6684	.6693	.6702	.6712
4.7	.6721	.6730	.6739	.6749	.6758	.6767	.6776	.6785	.6794	.6803
4.8	.6812	.6812	.6830	.6839	.6848	.6857	.6866	.6875	.6884	.6893
4.9	.6902	.6911	.6920	.6938	.6937	.6946	.6955	.6964	.6972	.6981
5.0	.6990	.6998	.7007	.7016	.7024	.7033	.7042	.7050	.7059	.7067
5.1	.7076	.7084	.7093	.7101	.7110	.7118	.7126	.7135	.7143	.7152
5.2	.7160	.7168	.7177	.7185	.7193	.7202	.7210	.7218	.7226	.7235
5.3	.7243	.7251	.7259	.7267	.7275	.7284	.7292	.7300	.7308	.7316
5.4	.7324	.7332	.7340	.7348	.7356	.7364	.7372	.7380	.7388	.7396
5.5	.7404	.7412	.7419	.7427	.7435	.7443	.7451	.7459	.7466	.7474
5.6	.7482	.7490	.7497	.7505	.7513	.7520	.7528	.7536	.7543	.7551
5.7	.7559	.7566	.7574	.7582	.7589	.7597	.7604	.7612	.7619	.7627
5.8	.7634	.7642	.7649	.7657	.7664	.7672	.7679	.7686	.7694	.7701
5.9	.7709	.7716	.7723	.7731	.7738	.7745	.7752	.7760	.7767	.7774

Table of Logarithms—Continued

	0	1	2	3	4	5	6	7	8	9
6.0	.7782	.7789	.7796	.7803	.7810	.7818	.7825	.7832	.7839	.7846
6.1	.7853	.7860	.7868	.7875	.7882	.7889	.7896	.7903	.7910	.7917
6.2	.7924	.7931	.7938	.7945	.7952	.7959	.7966	.7973	.7980	.7987
6.3	.7993	.8000	.8007	.8014	.8021	.8028	.8035	.8041	.8048	.8055
6.4	.8062	.8069	.8075	.8082	.8089	.8096	.8102	.8109	.8116	.8122
6.5	.8129	.8136	.8142	.8149	.8156	.8162	.8169	.8176	.8182	.8189
6.6	.8195	.8202	.8209	.8215	.8222	.8228	.8235	.8241	.8248	.8254
6.7	.8261	.8267	.8274	.8280	.8287	.8293	.8299	.8306	.8312	.8319
6.8	.8325	.8331	.8338	.8344	.8351	.8357	.8363	.8370	.8376	.8382
6.9	.8388	.8395	.8401	.8407	.8414	.8420	.8426	.8432	.8439	.8445
7.0	.8451	.8457	.8463	.8470	.8476	.8482	.8488	.8494	.8500	.8506
7.1	.8513	.8519	.8525	.8531	.8537	.8543	.8549	.8555	.8561	.8567
7.2	.8573	.8579	.8585	.8591	.8597	.8603	.8609	.8615	.8621	.8627
7.3	.8633	.8639	.8645	.8651	.8657	.8663	.8669	.8675	.8681	.8686
7.4	.8692	.8698	.8704	.8710	.8716	.8722	.8727	.8733	.8739	.8745
7.5	.8751	.8756	.8762	.8768	.8774	.8779	.8785	.8791	.8797	.8802
7.6	.8808	.8814	.8820	.8825	.8831	.8837	.8842	.8848	.8854	.8859
7.7	.8865	.8871	.8876	.8882	.8887	.8893	.8899	.8904	.8910	.8915
7.8	.8921	.8927	.8932	.8938	.8943	.8949	.8954	.8960	.8965	.8971
7.9	.8976	.8982	.8987	.8993	.8998	.9004	.9009	.9015	.9020	.9026
8.0	.9031	.9036	.9042	.9047	.9053	.9058	.9063	.9069	.9074	.9079
8.1	.9085	.9090	.9096	.9101	.9106	.9112	.9117	.9122	.9128	.9133
8.2	.9138	.9143	.9149	.9154	.9159	.9165	.9170	.9175	.9180	.9186
8.3	.9191	.9196	.9201	.9206	.9212	.9217	.9222	.9227	.9232	.9238
8.4	.9243	.9248	.9253	.9258	.9263	.9269	.9274	.9279	.9284	.9289
8.5	.9294	.9299	.9304	.9309	.9315	.9320	.9325	.9330	.9335	.9340
8.6	.9345	.9350	.9355	.9360	.9365	.9370	.9375	.9380	.9385	.9390
8.7	.9395	.9400	.9405	.9410	.9415	.9420	.9425	.9430	.9435	.9440
8.8	.9445	.9450	.9455	.9460	.9465	.9469	.9474	.9479	.9484	.9489
8.9	.9494	.9499	.9504	.9509	.9513	.9518	.9523	.9528	.9533	.9538
9.0	.9542	.9547	.9552	.9557	.9562	.9566	.9571	.9576	.9581	.9586
9.1	.9590	.9595	.9600	.9605	.9609	.9614	.9619	.9624	.9628	.9633
9.2	.9638	.9643	.9647	.9652	.9657	.9661	.9666	.9671	.9675	.9680
9.3	.9685	.9689	.9694	.9699	.9703	.9708	.9713	.9717	.9722	.9727
9.4	.9731	.9736	.9741	.9745	.9750	.9754	.9759	.9763	.9768	.9773
9.5	.9777	.9782	.9786	.9791	.9795	.9800	.9805	.9809	.9814	.9818
9.6	.9823	.9827	.9832	.9836	.9841	.9845	.9850	.9854	.9859	.9863
9.7	.9868	.9872	.9877	.9881	.9886	.9890	.9894	.9899	.9903	.9908
9.8	.9912	.9917	.9921	.9926	.9930	.9934	.9939	.9943	.9948	.9952
9.9	.9956	.9961	.9965	.9969	.9974	.9978	.9983	.9987	.9991	.9996

APPENDIX V

Suggested Locker Equipment

2 beakers, 30 or 50 cm^3
2 beakers, 100 cm^3
2 beakers, 250 cm^3
2 beakers, 400 cm^3
1 beaker, 600 cm^3
2 Erlenmeyer flasks, 25 or 50 cm^3
2 Erlenmeyer flasks, 125 cm^3
2 Erlenmeyer flasks, 250 cm^3
1 grad. cylinder, 10 cm^3
1 grad. cylinder, 25 or 50 cm^3
1 funnel, long or short stem
1 thermometer
2 watch glasses, 8 or 10 cm
1 crucible and cover, size #0
1 evaporating dish, small

2 medicine droppers
2 test tubes, 18 × 150 mm
8 test tubes, 13 × 100 mm
4 test tubes, 10 × 75 mm
1 test tube brush
1 file
1 spatula
1 test tube holder, wire
1 test tube rack
1 tongs
1 sponge
1 towel
1 plastic wash bottle
1 casserole, small